simplex*x*ity

OTHER NONFICTION *by* JEFFREY KLUGER

Splendid Solution: Jonas Salk and the Conquest of Polio

Journey Beyond Selene

The Apollo Adventure

Apollo 13 (written with Jim Lovell)

FICTION *by* JEFFREY KLUGER

Nacky Patcher and the Curse of the Dry-Land Boats

simplexity

Why Simple Things Become Complex (And How Complex Things Can Be Made Simple)

Jeffrey Kluger

HYPERION

New York

Library of Congress Cataloging-in-Publication Data is available upon request.

ISBN: 978-1-4013-0301-3
Paperback ISBN: 978-1-4013-0993-0

Hyperion books are available for special promotions and premiums. For details contact the HarperCollins Special Markets Department in the New York office at 212-207-7528, fax 212-207-7222, or email spsales@harpercollins.com.

Book design: Guenet Abraham
Charts and illustrations: Yann-Pablo Corminboeuf and Beatriz Tirado

FIRST TRADE PAPERBACK EDITION

10 9 8 7 6 5 4 3 2 1

To my family,
for keeping things
simple

Contents

simplexity

Prologue

London 1854

TO ANYONE PAYING ATTENTION ON the morning of August 29, the death of the little girl at 40 Broad Street did not seem like a terribly remarkable thing. Not many people in the London neighborhood where the girl had lived even knew her name—small children being something of an overstocked commodity in so overrun a place as Broad Street. Indeed, it's entirely possible no one outside the child's immediate family would have learned of her passing at all had it not been for the way she died: the violent intestinal spilling that was the unmistakable signature of cholera. Let the cholera bug alight even briefly and a single baby at a single address was not the only one who would die.

Cholera, the neighbors knew, would spread quickly, but it would also spread unpredictably. Unlike influenza—which could travel by air, seeming to bring down whole blocks in a single infectious sweep—cholera was choosy. It would tap some members of a household and spare others, strike one home on a block and leap over the next. The disease would sweep wide—of that there was no doubt—but in what direction and with what rhythm was impossible to predict.

The Broad Street case broke free just as fast as people feared it would. By the very next day, dozens of Londoners in the seventy-block area surrounding the girl's address had seized up with the illness. The following day, the plague claimed 100 more people, all in the same febrile quarter. The day after that 141 more victims were struck—nearly all of whom either quickly died or seemed certain to do so. An unusually ferocious case of cholera was clearly on the loose, and though for now it seemed content to busy itself with the people in a single working-class ward, it would not be contained there for long. If it was going to be controlled, it would have to be stopped at its source. John Snow, a forty-one-year-old physician who lived near the blighted quarter, thought he might know how.

Snow was already well acquainted with cholera. Five years earlier, a similar, if smaller, outbreak had struck near the London wharves, and he had been called in to treat the young sailor in the Bermondsey neighborhood who had been the first to fall ill. Snow was too late to save the man, and indeed was unable to save the next occupant of the sailor's modest rooming house quarters as well. This second man had moved in after the stricken man had died and soon grew sick himself. Snow examined what the men had in common. They had breathed the same air in the same small house, but so had the other residents and they had remained well. They had never shared a meal, nor was there any suggestion that they'd even dined in the same restaurant or public house. They'd used the same bed at different times, and it was always possible that the first man left behind a trace of disease that the second man had acquired, though subsequent residents of the room had remained healthy. Then too it was always possible that the two dead men had drunk the same water.

London neighborhoods had for some time been well plumbed, outfitted with public pumps that provided unlimited supplies of water for washing, cooking, and drinking. The Bermondsey neighborhood where the sailor had died was no exception, nor was the seventy-block area where the Broad Street cholera was now raging. The entire quarter had no fewer than seven pumps, as well as two more just outside its borders.

As was the custom in London, people grew picky about the pumps they used, not always preferring the one closest to their homes, but the one that produced water whose taste was most to their liking. One member of a household might simply step to the corner to draw a dipper of water, while another might walk several blocks away. If there was any disease teeming in the waters that fed one pump and not another, it would be the people drinking from the fouled one who would get sick first. And the London plumbing had plenty of reason to grow foul.

As many pumps as there were in all the London neighborhoods, there were far more cesspools—two hundred thousand of them by the middle of the nineteenth century. There was a time when the pools were kept surprisingly clean, their contents regularly drawn out as a handy source of fertilizer for the farms still located in and around the city. The money that the sewage sales brought in was used to maintain the brickwork that lined and sealed the pools. Over time, however, the farms migrated farther and farther from the city and the farmers purchased their fertilizer deeper and deeper in the country. With no one regularly draining the city cesspools and no funds to maintain the immersed brickwork, the masonry seals began to fail and the contents of the pools began to seep into the nearby water supply. Snow suspected that, somewhere in the sickly seventy-block quarter, there was a diseased pool fouling a pump. Shut down that pump and you might stop the contagion.

On September 2, four days after the epidemic began, Snow took to the streets. As dozens more people continued to fall ill every few hours, he began calling on the houses of those already dead and those freshly stricken to find out how many people lived in each home and which ones had drunk their water from which neighborhood pumps. The mother who lost her daughter at 40 Broad Street usually used the closest pump, which was just outside her door and only three feet from a neighborhood sewage pool. She didn't regularly drink the water that flowed from it, but she did use it to rinse and soak her daughter's diapers. Other people who lived near her also frequented the Broad Street pump, but still others preferred the ones at Bridle or Warwick or Rupert Streets. Snow inked all the names and addresses of all of the people he interviewed into a ledger and then drew up a map, with spidery lines connecting the houses of the well and the stricken to the site of the pumps that provided their water. In home after home, he noticed that while the odd victim or two couldn't remember where they'd drawn their water, almost every one of the other people felled by the disease had made it a point to seek out Broad Street water.

Over the week, as the death toll crossed four hundred and then five hundred, Snow continued his canvassing, ranging farther and farther from Broad Street, out to the cobbly alleys and avenues where the disease was its weakest. He stopped in at a neighborhood jail where five hundred inmates were kept in frightful conditions, yet only five had contracted cholera—due, perhaps, to the fact that the building had its own well. He stopped in at a nearby brewery, where workers were given a generous ration of fresh beer each day, sparing them the need to sip much from any of the pumps. No one there had gotten ill.

Finally, on September 7, when the death toll had reached nearly six hundred, Snow demanded and received an audience with the Board of Guardians of St. James Parish at the Vestry Hall near his home. Ledger and map in hand, he appeared before the committee and told them plainly that the Broad Street pump was the source and center of the recent contagion. Disable it and the plague would be stayed. The guardians demurred: One pump causing so much suffering in so broad a range? they asked. Yes, Snow explained, and handed over his evidence. The men studied the material and remained unconvinced. Eventually, however—perhaps to appease the anxious doctor, perhaps because there were so many other functioning pumps in the ward that the Broad Street residents would not be too inconvenienced if theirs was shut down—the guardians agreed to Snow's recommendation. The next day, a workman was sent to the offending pump, heavy mallet in hand. He took several clanging swings at the fittings holding the handle in place. The metal peg popped free and the handle itself clattered to the street, rendering the pump useless.

That day, only fourteen people fell newly ill. The following day, the number dropped further still. Within a few more days, the dying stopped and the disease was beat.

John Snow knew nothing of the modern science of epidemiology—no surprise since it was a science that did not yet even exist. But he knew about diseases and he knew two things about them: Plagues were fantastically complex things—with the illness working myriad horrors in the body and spreading across the landscape in myriad ways. But diseases moved through simple choke points too—one person's handkerchief, one person's handshake, one handle on one fouled pump—all of them bottlenecks in the pathogen's flow. Seal off even one of those and you could stop the disease cold. The complex illness could collide hard with the simple fix. And on this day at least, it was the simple fix that held its ground.

TO ANYONE PAYING attention, it's clear something unremarkable is going to happen when the M. Coy bookshop in Seattle opens its doors for business in the morning. It's a big, roomy place, but an inconspicuous one too, a store you have to look for along the brick-paved stretch of Pine Street that spills down to the Farmer's Market. Those who have patronized the place know it's been here for a while now, eighteen years or more, run all that time by business partners Michael Coy and Michael Brasky.

Coy and Brasky keep careful track of their books, stocking about fifteen thousand titles at any one time. Counting multiple copies of popular books, that comes out to about twenty thousand volumes in all, most of which appear to be on display. Twenty thousand books is an awful lot if you're two men running a single shop, though to a vast book chain with outlets around the world, that same twenty thousand would be a mere rounding error—perhaps the amount left in loading dock crates at the end of the day, to be unpacked in the morning after the early shift arrives.

When the doors of the shop on Pine Street open a little before 10 A.M., no one comes in at first. Then a pair of browsers arrive, then a single man, then a group of three—and Coy and Brasky go to work. Following the meandering customers with their eyes, they note who stops where, who pauses at which shelf, who picks up a book and puts it down quickly, and who picks up another one and gives it more attention. They notice the shoppers who keep glancing at their watches—the ones likeliest to have stopped in to burn ten minutes on their way to somewhere else. They notice those who move more slowly—the ones for whom the store was a destination, not a transit point. They notice which customers have been here before and which ones are newcomers.

The proprietors gather those stray scraps of information and, as other customers wander in, go panning for more. A youngish woman stops at a wall of children's books. If she's wearing a wedding ring she may be a mother familiar with what's on offer. If there's no ring, maybe she's an aunt or a family friend who needs some help. The owners spot connections between two books that attract a single shopper—a period novel set in an artist's studio and a coffee table book about famous painters. Maybe something there, maybe not. They remember which books repeat customers have bought in the past, recall other customers who have bought the same book, and look for still more books both customers might enjoy. They discard information that would only confuse things. People who ask for books to be gift-wrapped don't count. They're buying for someone else's tastes, not their own. But a year from now, when the same birthday or anniversary comes around and the same customer returns, that information might become useful.

Coy and Brasky stir all these thoughts together, adding this or that odd detail or this or that strategy from a lifetime spent bookselling, and do it all instantly—tiny synapses flashing tiny signals among billions of neurons, summoming up volumes of experiences stored all over the brain. Then they approach a shopper with the most important thing they have to offer: a recommendation, a suggestion for a single one of the twenty thousand volumes they keep on their shelves.

Sometimes the customers don't like the idea, but more often they do. Between sixty and one hundred times a day the cash register rings and a sale is recorded—meaning that more than half a million books have gone out the door over the course of eighteen years. "It's called book sense," says Brasky. "You get a feel for what sells, for what people will like. Then you offer it to them."

In another part of Seattle, not far from the store, people are acquiring book sense too. But they're not doing it in a one-story shop on a brick-lined street. They're doing it in the 1927 art deco tower that was once a private hospital and is now the worldwide headquarters of Amazon.com.

Amazon has a few more books than Brasky and Coy. They don't say precisely how many, but they acknowledge that "tens of millions" starts to get at it. They have twenty-one warehouses around the world to store that mountain of merchandise, along with all the other products—movies, clothing, jewelry, toys, cameras, tools, sports equipment, kitchen supplies—their customers buy. They also have big computers, a lot of big computers, but they keep even quieter about those, saying nothing about how many they own, how the network is set up, or especially about where they keep them all. "They'd be a target," company spokesman Andrew Herdener says simply, leaving the matter at that.

The Amazon folks are plenty open about what their

computers can do, however. Log onto the site and you've walked into their store. As you do, the computers keep an eye on you. They see where you click, they see where you linger. They know if you start to buy something but decide against it before you complete the sale. They remember what you've bought before and what other people like you have bought. They cross-index those millions and millions of customers and tens of millions of sales and look for what overlaps. They too find the hidden connections in seemingly unconnected books—the Tuscany tucked deep in a novel you bought and the Mediterranean travel guide you consider the next week. They too know to ignore the confounding junk. Throw a Harry Potter book in with your purchase and the computer's memory will likely throw it right out. Everyone buys those so they tell you nothing.

The computers are there for the same reason Brasky and Coy are—to make recommendations, to offer a single book or an entire reading list that you in particular would be likely to enjoy. They sift uncountable billions of data bits too. However, the computers don't do it in a fourteen-hundred-gram, walnut-wrinkled mass of neural tissue and myelin, but in vast mainframes in hidden hangars concealed around the world.

"The bookstore owner's brain is not as large as everybody's collective brain," says Scott Ruthfield, an Amazon software manager. "We try to get humans out of the equation. Humans think they're experts and try to guess the patterns. But they're usually wrong."

The likes of Coy and Brasky don't think they're wrong. And they're not just in the equation, they are the equation.

What the Amazon computers do is by any measure remarkable—just as remarkable as the celebrated chess-playing computers that can humble even the world's most gifted human players. But in both those cases, the machines aren't parsing data as much as steam-shoveling it, digging up vast masses of information and pulverizing it finer and finer until they come up with a single relevant grain. The likes of Brasky and Coy do the very same thing, but finely, instantly, microsurgically. It's easy to say that one of the two approaches is more complex than the other; it's a lot harder to say which one.

COMPLEXITY, AS ANY scientist will tell you, is a slippery idea, one that defies almost any effort to hold it down and pin it in place. Things that seem complicated can be preposterously simple; things that seem simple can be dizzyingly complex. A manufacturing plant—with its clanking machinery filling acres and acres of hangar-sized buildings—may be far less complicated than a houseplant, with its microhydraulics and fine-tuned metabolism and dense schematic of nucleic acids. A colony of garden ants may similarly be more elaborate than a community of people. A sentence may be richer than a book, a couplet more complicated than a song, a hobby shop harder to run than a corporation.

Human beings are not wired to look at matters that way. We're suckers for scale. Things that last for a long time impress us more than things that don't, things that scare us by their sheer size strike us more than things we dwarf. We grow hushed at, say, a star, and we shrug at, say, a guppy. And why not? A guppy is cheap, fungible, eminently disposable, a barely conscious clump of proteins that coalesce into a bigger clump, swim about for a few months, and then expire entirely unremarked upon, quickly decomposing into the raw chemicals that produced them in the first place. A star roars and burns across epochs, birthing planets, consuming moons, sending showers of energy to the limits of its galaxy.

Yet the guppy is where the magic lies. A star, after all, is just a furnace, a vulgar cosmic engine made up of three layers of gases that slam hydrogen atoms together into helium, release a little light and fire in the process, and achieve nothing more than that. It may last for billions of years, but to what animating end? A guppy, by contrast, is a symphony of systems—circulatory, skeletal, optical, neurological, hemotological, metabolic, auditory, respiratory, olfactory, enzymatic, reproductive, biomechanical, behavioral, social. Its systems are assembled from cells; its cells have subsystems; the subsystems have subsystems. And if so elegant an organism lives for no more than a handful of months, what of it? Admire the fireball star if you must, but it's the guppy we ought to praise.

Across all the disciplines—chemistry, physics, astronomy, biology, economics, sociology, psychology, politics, even the arts—investigators are making similar discoveries, tilting the prism of complexity in new directions and seeing the light spill out in all manner of unexpected ways. Listen to an economist talk about a pencil. There is nothing to recommend a pencil as a remotely interesting thing.

Its design is crude, its materials humble. It operates—to the extent that the word can be used at all—by an act no more sophisticated than scraping. You could do as much with a scorched bit of charcoal and a flat stone wall.

And yet a pencil, by almost any measure, is an exquisitely complex thing. Somewhere in a harvested forest was the stand of cedar trees that gave up their wood to provide the pencil's dowel. Somewhere in the Jamaican interior is the bauxite mine that provided the raw material for its little aluminum sleeve. Somewhere in the coal belt is the mine that provided the lump carbon for the graphite. Still elsewhere is the lab where raw polymers were cooked up into rubbery erasers. And to those places that provided those things streamed still other things—the smelting ovens for the metal plants, the autoclaves for the rubber labs, the blades for the sawmills, the backhoes for the carbon mines, the cotton to dress the lab workers, the bacon to feed the lumberjacks, the paymasters and truck drivers and box packers and shipping managers to keep all of the operations humming. A vast industrial machine rises up, switches on, and at its far end, spits out . . . a pencil, arguably one of the most complicated objects in the world.

Psychology is waking up to complexity too. The six-child family may be an impossibly complicated thing with all of the lunches to be made, clothes to be bought, and homework assignments to be checked. But it may be the one-child family, in which a smaller number of relationships play out in a more nuanced way, that is the truly complex one. The single small mind of a single small child may be similarly more layered than it seems. A toddler walking down a sidewalk holding a toy in one hand unthinkingly switches it to the other so he can reach up and take hold of his mother before he crosses a street. But that apparently simple act is wonderfully complex, requiring the child to know a lot of things about streets (they're dangerous), his mother (she expects him to hold on), his grip (he can't hold both his mother and his toy at the same time), and gravity (let go of the toy without changing hands first and it will fall to the ground).

Political science plays the simplicity-complexity game too. Vast ideologies turn on simple ideas—Nazism on hate, democracy on consent of the governed, communism on the selflessness of the human spirit and the perfectibility of the marketplace. One idea may be ugly, one inspired, one badly flawed, but they're not hard to describe. Equally surprising and counterintuitive is the science of the stock market, where millions of blind and self-interested trades somehow settle on a fair value for tens of thousands of different companies; the science of crowds, in which terrified or rage-blind mobs do cunning and adaptive things, usually without being aware they're doing them at all; geology, in which the simple, chemical rules of rocks are increasingly seen to mirror the complex, chemical rules of biology; manufacturing and marketing, in which vast operations like industrial farms are humbled in the face of the mere greengrocer, who doesn't handle just a handful of products but thousands, all of which have to be shown and sold or moved and junked within the knife-edge margin that spells the difference between profitability and bankruptcy; art, in which jazz is no longer seen as mere music, but a living example of a complex adaptive system, and Jackson Pollock abides by not merely the laws of artful abstraction, but the laws of fractal geometry. Even presidential politics are remarkable for the complexity tucked in their simplicity, as whole administrations stand or fall or aren't born at all based on nothing more than a few chads in Florida, a piece of tape on an office door in Washington, a helicopter burning in an Iranian desert.

Trying to distill all of this down to a working definition of just what simplicity is and just what complexity is has always been difficult. Perhaps simple things are those things that create order in their environments, the way, say, modern design strips the clutter from a building. But complex things create order too, the way the cogs and springs and flywheels of a watch yield a single, simple piece of data—the time—and that piece of data in turn runs the world. Perhaps it's simple things that create disorder then, the way an underregulated economy can let markets and monopolies run wild. Or maybe it's complex things that are to blame, the way overregulating human affairs can produce a Byzantine mess like the tax code. We call some software simple because it's easy to use, never mind the fact that it relies on a heiroglyph of code that's numbing in its elaborateness. We call older computers complex because they were hard to use, never mind the fact that by the standards of today's machines they're flat-out crude. Is free-form dance comparatively simple thanks to its lack of rigid rules, or does that very liquidity make it much harder to learn? Does a committee struggling over a problem lack the straightforward insight of a single wise mind, or is the wise mind blinkered by its biases, while the committee—with its clarifying tensions—is a better, ultimately less complex arbiter?

IT'S NOT OUR fault that we lack a natural ability to tease these things apart. The human brain is a real-time machine, one that's designed to scan constantly for input, for clues to the next meaningful thing that's about to happen in our world, and then assemble that information quickly into impressions and actions. That kind of cognition may have been the only way for the species to survive in the wild, but it can mislead us now, causing us to overfocus on the most conspicuous features of a thing and be struck—or confused—by that quality. Thus, we are confused by beauty, by speed, by big numbers, by small numbers, by our own fear, by wealth, by eloquence, by size, by success, by death, by the unfathomability of life itself. There is a taxonomy of things that fool us every day and, in so doing, help the complex masquerade as the simple, and the simple parade itself as complex.

Distinguishing between the two is not easy, and complexity science doesn't pretend to have all—or even most—of the answers. By nearly any standard it's a young discipline, one just finding its feet. Its insights can be keen and its ideas compelling, but for every theory it proposes, there can be a countertheory; for every QED, a caveat. The study of complexity is less similar to a settled field like geometry or statistics than it is to microbiology in its early days or genetics shortly after the discovery of the double helix—a science only beginning to stir, one that doesn't yet offer up hard conclusions and proofs, but rather, an exciting new way to go in search of them. Being present at the birth of such a field can be a good deal more thrilling than merely studying its findings after its work is done. It's a rare thing to witness such a hinge-point in the history of science—and, as a walking tour of some of the disciplines of complexity shows, an awful lot of fun too.

CHAPTER ONE

Why is the stock market so hard to predict?

Confused by Everyone Else

WHAT MIGHT BE THE THREE MOST expensive words ever spoken were uttered on October 15, 1987. Few people suspected the impact of the words at the time, their seeming blandness entirely belying their power. But the words weren't chosen to be bland. In the coded vocabulary of the diplomatic world, they were actually chosen for their bite. It was just that nobody expected they'd bite so hard.

It was then-Secretary of the Treasury James Baker III who did the speaking that day, not long after returning to

Washington from a meeting in Bonn with his West German counterpart. Baker had good reason to be in that particular country meeting with that particular minister. The German economy, which had long been the strongman of Europe, had pulled a charley horse of late, with the mark stumbling significantly and the ability of the German people to buy pricey American goods falling with it. If those goods stayed on the shelves when the U.S. economy itself was already struggling under the twin loads of growing trade and budget deficits, the strain could be just enough to trigger a recession. This did not please Washington, and Baker had gone to Bonn in part to carry the message that he—to say nothing of President Reagan—would appreciate it if the Germans would lower their interest rates and juice the mark a bit, so that their consumers' buying power would rise commensurately. The German finance minister made it clear that he would lower his interest rates when he wanted to, and this was not that time. In fact, he might even raise them a tick, as he had on other occasions recently.

Baker was unhappy with this response—"not particularly pleased" was the three-word way he described it—and hinted darkly that if German rates did not drop, he might have no choice but to free up the dollar a little and allow it to drift against the mark. This might level things out in the short term, but it would increase the risk that the cold both economies had caught of late would turn into a full-blown flu. It was a good threat and a well-phrased one, precisely the kind of diplomatic sempahoring the situation called for. It no doubt did the job in Bonn, but more than just the Germans got nervous.

By the next morning, Friday, October 16, word of Baker's warning seeped through the investment community. When it did, investors decided they wanted no part of the looming catfight over currency. Markets had spent the Reagan era on a near-seven-year climb, and now seemed like a good time to cash out some winnings and head for the exits. That day, the Dow Jones average fell 108.36 points, a 4 percent loss that was a serious blow even for an index that had opened the day above 2,355. The plunge suggested only worse things to come, and a Friday was either a very good time or a very bad time for the first upheaval to have happened—good because traders would have the weekend to collect themselves and lower the temperature before the market opened again on Monday; bad because they might instead use that two-day respite as a time to worry and stew. On Monday, October 19, at 9:30 A.M., it quickly became clear which course they had chosen.

In the first thirty minutes of trading on that nervous morning, 50 million shares were sold, knocking fully sixty-seven points off the Dow, or 2.8 percent of its value. This was exactly the kind of early bleeding analysts had spent the weekend dreading, and at the news that it had begun, more investors—the majority of them with nothing more than small personal portfolios—began to unload their holdings as well. Within another half hour, a total of 101 points had been slashed from the Dow, 4.2 percent of its value at the opening bell, and 140 million shares had been sloughed off—or about the typical trading volume for an entire day. By eleven o'clock yet another hundred points were shed and tens of millions more shares unloaded, and that didn't tell the entire story. The Dow ticker, unable to process the flood of sales, was running nearly twenty minutes behind. There were surely untold millions of shucked shares hidden in that third of an hour.

Now the global markets' nightmare scenario—a blind stampede—began. By noon, the market losses totaled 240 points, a 10 percent loss in just a matter of hours. Spectators began to jam the gallery of the New York Stock Exchange, watching both the bloodletting below and the news from overseas streaming in on the big boards. Around the world, those boards showed, the markets of Europe, the Pacific, and particularly Japan were also in chaos, losing value almost as fast as the American one. What's more, by now, computers had joined the frenzy the investors had started, with programmed trades designed to dump dying stocks when their prices fell below certain levels doing just that—behaving no more rationally than the panic-prone people who had written the hair-trigger software. The more the computers tossed their holdings overboard, the more the individual investors followed suit, making it likelier that the machines would jettison still more.

By 2:00 P.M., the Dow free-fell through the two-thousand mark, and with that psychological floor no longer holding prices up, the plunge accelerated even further. By 2:15 the loss stood at three hundred points, or 12.7 percent off the market's opening value. Before 3:00 it was at four hundred, a sickening 17 percent loss. With the market ticker now running a full 111 minutes behind—as good as if it weren't running at all—no one doubted that the crash was reaching some kind of terminal velocity.

The only thing that would stop it would be the hard pavement of the end of the trading day, which at long last arrived at 4:00. When the closing hammer finally fell, the Dow was in a shambles, having lost a record 508 points—or 23 percent of its value—on an unprecedented 604 million traded shares. (At the thirteen thousand mark the Dow broke for the first time in early 2007, that would be the equivalent of a one-day loss of just under three thousand points.) In just that six and a half hours, half a trillion dollars in American wealth had been incinerated. Overseas, the Tokyo exchange shed 57 trillion yen, or 400 billion dollars; the London market lost 94 billion pounds, or 140 billion dollars. The French, German, Canadian, Australian, and Mexican markets all lost between 9 and 30 percent of their value. Four days earlier, the American treasury secretary had spoken, and at least partly in response to his handful of words, the world's financial markets had set themselves on fire.

Picking through the wreckage in the days that followed, analysts could only nod at the irrational—and predictable—mess that investors had made of their own wealth. Even with the growing deficits and the other market stresses—even with Baker's coolly lethal comment—nothing smart or sensible had taken place that day. But smart and sensible forces had not been at work. Market forces had been. And when it comes to simple, there's little to compare to those.

Never mind what you think about the exquisitely complex organism that is the world's financial markets. Never mind the hundreds of millions of thoughtful investors and their billions of well-considered trades. For every market analyst who sees traders as the informed and educated people they surely can be, there are scientists who see them another way entirely: as atoms in a box, billiard balls on a table, unthinking actors who obey not so much the laws of economics as the laws of physics. The things that result from those actions may be undeniably extraordinary—the creation or destruction of trillions of dollars of wealth in a matter of hours—but down at the fine-grained level at which the transactions are made, the players themselves can be remarkably simple things.

Investors react not so much to variables that are in their interests, but, oddly, to those that are in everyone else's interests. When the tide of the market shifts, most of us shift with it; when it flows back the other way, we do the same. We like to think we're informed by trends, but often as not, we're simply fooled by them—snookered by what everyone else is doing into concluding that we ought to do the same.

"Economic models always begin with the assumption of perfect rationality, of a universe of logical people all doing what they can to master their utility," says economist and ecologist J. Doyne Farmer, a former fellow at Los Alamos National Laboratories and now a resident faculty member at the Santa Fe Institute (SFI) in New Mexico, a think tank and research center devoted to the study of complexity. "Physicists studying economics begin with the assumption that people can't think."

"The term we use is zero-intelligence investors," says John Miller, an economist at Carnegie Mellon University specializing in complex adaptive social systems, and another member of the SFI faculty. "It's not a term that flatters the investors, but it does go a long way to describing what goes on."

If there's any group that has the credibility to make such provocative claims it's the Santa Fe Institute. And if there's anyone equipped to lead SFI's work, it's physicist Murray Gell-Mann.

Gell-Mann won a Nobel prize in 1969, the year he turned forty, for being the first investigator to sort out the confusion quantum theorists confronted when they studied the goings-on inside the atom. Since the late nineteenth century, physicists had been coming to the conclusion that the supposedly irreducible atom was not the last word on the structure of matter, and over time they identified a gnat-swarm of some one hundred subatomic particles tucked inside the atomic nucleus. Nobody, however, had been able to make any order of them all until Gell-Mann came along. He developed a sturdy new theory that sorted the particles into several different species of what are now known as quarks, all held together by forces he called gluons. The work gave instant clarity to an entire field and remains the core of particle theory into the twenty-first century.

Fifteen years after winning his Prize, Gell-Mann was approached by a small confederation of scientists wanting to know if he was interested in helping to found a nonprofit research center in the high desert of New Mexico to study the twin puzzles of what makes something simple and what makes it complex. Gell-Mann instantly took to the challenge, reckoning that having already brought discipline to the chaotic world of quarks, he might enjoy doing the same with the equally unruly swirl of specialties and subspecialties that makes up scientific theory. In 1984, the Santa Fe Institute was created, with Gell-Mann as its cofounder. Today, SFI is a hive of more than one hundred resident faculty members, postdoctorate fellows, visiting scholars, and other researchers, studying the internal clockwork of dozens of fields. Gell-Mann is still there, a philosopher father of sorts, and one who continues to give the group its gravitational center.

To the outsider, the precise function of SFI is not easy to grasp. The group's mission statement, printed in its annual reports and announced on its website, is itself hardly a model of simplicity, defining the Institute as a "multidisciplinary collaboration in pursuit of understanding the common themes that arise in natural, artificial, and social systems." To the extent that that's comprehensible at all, it sounds an awful lot like a think tank, where learned people just sit about and, well, think. And after a fashion, that's what the SFI scientists do. There is, however, more to it than that. In the cluster of the institute's interconnected buildings, or "pods" as the complexity scientists call them, all manner of things go on.

Over here is an ecosystems biologist looking at the elaborate food webs within a single small pond, with ninety-two different species eating or getting eaten in more than nine hundred different predator-and-prey combinations. Knock one organism out and you may not upset things too much; knock out a handful and the web starts to shake. Pick the one life form on which all of the others perch—the so-called keystone species—and the entire system can cascade. Think the computer models that explain all that don't have implications for human food sources, not to mention human social systems, economies, and even physical structures? Think again.

Elsewhere, someone is investigating the network of relationships in communities, office places, and even monkey troops, looking for the larger rules that can explain how information gets disseminated and ideas get exchanged, with implications for everything from politics to public health. Still elsewhere, physicists are studying the metabolism of animals, looking for the patterns that will reveal more about the life and death cycles not only of living things, but of institutions, cities, and even nations. All of these insights may reach their first audiences in obscure journals that nobody but other academics read. But all of them will then work their way through the knowledge web and find applications in the real world—much in the same way that the ideology that's driven the political right in the last twenty years was born in think tanks, as was much of what drove the political left in the decades before that.

Complexity, as both a scientific and social concept, is arguably more powerful than anything that comes out of a political institute. The first trick, however, is defining it. One of Gell-Mann's favorite ways of deciding whether something is simple or complex is to ask a decidedly unscientific question: How hard is it to describe the thing you're trying to understand? A rock? Easy. A car? Harder. The physics behind a supercollider or the structure of a Mozart opera? That'll take a while. "You start with minimum description length as your first inquiry," he says. "The shorter it is, the simpler the thing is likely to be. In most cases, you don't even have to take your language from scientific discourse. It can come from ordinary speech."

But description depends on context and context changes everything. Explain the protein coat of a newly discovered virus to a molecular biologist and you're in and out in a sentence or two. Explain it to someone utterly unfamiliar with virology and you'll be at it a lot longer. A football player or a soldier can similarly understand a new play or a battle plan with just a few chalk strokes; the virologist would be utterly flummoxed. "Imagine an anthropologist approaching a civilization with which he shares a common language, but which is naïve of any culture outside of its own," Gell-Mann says. "Now imagine trying to explain to that community a tax-managed mutual fund. What do you think the preamble to that explanation would be?"

Description length, however, has its limits. While it's a handy place to start thinking about complexity, it's naïve in its own way. The problem is that it begins with an assumption of a clean and consistent line, with simplicity and short descriptions at one end and complexity and long descriptions at the other. A clean line, however, doesn't really capture things as much as a somewhat messier arc does.

Complexity scientists like to talk about the ideas of pure chaos and pure robustness—and both are exceedingly simple things. An empty room pumped full of air molecules may not be a particularly interesting place, but it is an extraordinarily active one, with the molecules swirling in all directions at once, dispersing chaotically to every possible crack and corner. On the other hand, a lump of carbon chilled to what scientists call absolute zero—or the point at which molecular motion is the slowest it can possibly be—is neither interesting nor active. The carbon is exceedingly static, or robust, as complexity researchers call it; the room is exceedingly chaotic. What neither of them is, however, is complex, offering only spinning disorder at one end and flash frozen order at the other.

Where you'd find real complexity would be somewhere between those two states, the point at which the molecules begin to climb from disorder, sorting themselves into something interesting and organized—a horse, a car, a com-

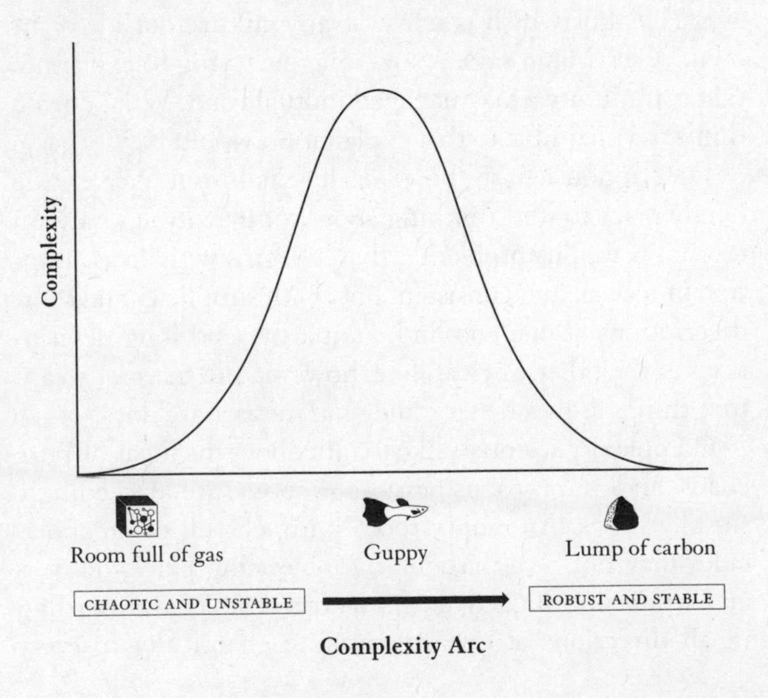

Complexity Arc

munications satellite—but catching themselves before they descend down the other side of the complexity hill, sliding into something hard and lumpish and fixed. The more precisely the object can balance at the pinnacle of that arc, the more complex it is. "It's the region between order and disorder that gives you complexity," says Gell-Mann, "not the order and disorder at the ends."

Things, however, are not as straightforward as even that explanation suggests, since any one system is not necessarily composed of just one point on the arc. Often, many different points come into play and the question of complexity turns on which one you choose. A foot-long length of copper pipe might be nearly as static as the frozen carbon. But step back and take in the vast, arterial array of skyscraper plumbing of which it is just a part and things look a lot more complicated. Drill deep inside the copper and consider the subatomic universe within its atoms and the picture becomes more complicated still.

The same kind of multitiered complexity holds true for human behavior. Consider a handshake, one of the simplest and most retail of social exchanges. Now close in and consider the things that allow it to happen at all—the neural firings, the tactile feedback, the fine control of muscle fibers, the unerring depth perception that permits two hands controlled by two people to arrive in the same spot at the same moment, even the optics that allow two pairs of eyes to meet and agree on the shake to begin with. Now pull back and consider what the little ritual arises from in the first place: the long, elaborate human tradition of head bobs and bows and curtsies and waves and hat tippings and by-your-leaves—all the social cues and deference gestures that have evolved over the millennia and allow us to live together peacably. Viewed big or viewed small, a handshake is an awful lot more than the unremarkable habit it seems to be.

Any system—chemical, physical, cultural, fiscal—must be seen at all of these levels before you can begin to make a real determination about whether it can truly be called complex. "Ask me why I forgot my keys this morning and the answer might be simply that my mind was on something else," says SFI neuroscientist Chris Wood. "Ask me about the calcium channels in my brain that drive remembering or forgetting and you're asking a much harder question."

The extent to which the stock market can be called complex then depends on where it falls along these scales and how it satisfies these definitions. Looking at things that way, there are a lot of people who would argue that the market shakes out as a pretty simple thing. Few places capture that idea better than the lab of economist Blake LeBaron at Brandeis University.

LeBaron can speak about the dynamics of markets with unusual authority in part because, unlike other scientists and commentators, he has an entire stock exchange he can call his own. LeBaron's exchange, to be sure, does not exist in the real world, but only in the world of his computers. Still, it's a wonderfully dynamic place. Over the years, he's developed algorithms that allow him to simulate any kind of market he wants—bull, bear, static, active, mixed—and then release simulated investors into that environment to see how they behave.

The universe he's built is admittedly limited: For one thing, there are only about a thousand imagined traders at work there at any time. For another, it moves exceedingly fast; market cycles that take months or years to play out in the real world require only a day or two in the computers. But on the whole, the artificial traders behave precisely like the real ones, which is to say that they almost never show much imagination. Watching his screens like a fish-keeper observing his tanks, LeBaron knows he will almost always see a familiar pattern play out.

"At the beginning of a run," he says, "all the traders just wobble around a bit, looking for guidance. Then someone will try a new strategy and do quite well at it and the others will notice. Pretty soon, a few more are trying it and it starts to get popular, like a clothing fad. Ultimately everyone starts to converge on that strategy and it dominates the market, precisely the way real markets behave."

But such high-profit homogeneity works only so long. If everyone is chasing the same dollar in the same way, it takes only a few players to cash out before all of the other shares start to lose value. This is the classic bubble-popping phenomenon, one that's familiar to all investors. It happened most recently in 2001, when the entire sector of Internet shares crashed, taking down the larger markets with it. It happened in 1998, when the spectacularly successful hedge fund Long-Term Capital Management revolutionized the way the hedge fund world operated—in part by brilliantly managing its investment in government bonds—generating such massive profits that everyone around them started imitating the strategy. When the fund's own fortunes started to falter, however, there was no one left to buy in, since everyone involved in that kind of speculating was already spoken for. It happened decades earlier too, in the 1970s, with the plunge of the so-called Nifty Fifty, a group of fifty blue chip stocks like Polaroid, Halliburton, Philip Morris, and Avon that were said to be so stable all you had to do was own them and you'd never have to worry about market uncertainty again. It was true for a while, until the millions of traders wanting a piece of the sure thing drove the prices of those stocks so unrealistically high they could do nothing else but crash.

"The market first becomes too stable, too robust," says LeBaron. "Then it collapses into instability." In other words, starting off at the frozen carbon end of Gell-Mann's arc, the market simply leaps to the roomful-of-gas end, pausing only comparatively briefly at the true complexity in between. And the majority of investors make the leap along with it.

The comparison of LeBaron to fishkeeper is actually an apt one, not only because it describes how he conducts his work, but also because it suggests the limits of that work. Nobody pretends that an aquarium is anything like a complete model of the complexities of a living river or sea. Rather, it offers only a broad-stroke idea of how the real aquatic world behaves. Still, the observant aquarist can learn a lot about chemistry, botany, and animal behavior by watching what takes place on the other side of the glass. So, too, can the market modeler get a clearer understanding of the dynamics of the trading floor by studying the computer screens.

Nonetheless, some researchers actually have turned to the behavior of fish to learn more about how traders conduct themselves. Simon Levin, a biologist and ecologist at Princeton University, recently worked with a team of three co-investigators to study how schools of fish manage the fluid, almost balletic movements they do without any evident alpha fish leading the way—a behavior that can be mirrored in the markets. The researchers began their work by studying the speed, motion, and population size of schools of real fish and then developed software models that replicated that behavior in the computer. They then teased those models apart, separating the strands of behavior fish by fish.

As a rule, they found that, in order for a traveling fish school to remain cohesive, all of the individuals must maintain a position no more or no less than about a single body length from every individual around them. Get too close and you collide and compete. Drift too far apart and you begin to stand out—never a good idea when predators lurk, looking for easy pickings at the fringe of the group. The most important thing the fish thus keep in mind—to the extent that they keep anything in mind at all—is not where they're going or who's leading them there, but making sure they don't fall out of ranks along the way.

Of course, fish on the move are inevitably heading somewhere deliberate—toward food or a spawning ground, or away from predators—and they can't just be getting there by guesswork. So who's in charge of the motion? Levin and his colleagues found that the schools do require at least a few leaders—"informed individuals" as they call them—but not many. In an average-sized school, it takes only about 5 percent of the members to know the proper route and set out in that direction for the other 95 percent to follow. What's more, as the school grows, that leadership share actually shrinks, so that as few as one in a hundred fish need have any idea what the goal is and yet they still lead all the others.

"Among aggregating animals like fish and like us," says Levin, "it takes only a few individuals to set off all kinds of complex behavior. The collective dynamics of a small minority in any herding community can trigger a cascade."

Simplifying things even further, in some cases it doesn't take human brains or even fish brains to set off such chain events. Indeed, it doesn't take something living at all. Often, the markets mimic the entirely nonsentient behavior of physics.

J. Doyne Farmer, the one-time Los Alamos researcher and current SFI faculty member, wondered if it might be possible to predict the volatility of markets by modifying an equation physicists use to determine the pressure of a gas in a box, beginning with its volume and temperature and factoring those values through a mathematical constant that's a bit like pi. That arcane bit of arithmetic can be pretty important if you're, say, designing an engine and want to know if you've calculated your pressures right without having to build the whole thing and let it blow up on you. Markets, in their own way, can blow up too, and it would be awfully nice to know if that's going to happen before it does.

Farmer devised such an equation, an exceedingly clean piece of ciphering that combines just three variables: market orders (when a stock is bought at whatever price it's selling for at the moment), limit orders (when an order is placed for the stock but the purchase is not actually made unless the price per share falls below a certain point), and the frequency with which some of those orders are canceled. The market orders increase the volatility, the limit orders decrease the volatility, both behaving like gas atoms bouncing about and colliding with one another. The cancellation of orders serves the same function as radioactive decay, taking some atoms out of the mix altogether. Factor all of those out the right way, Farmer found, and volatility can be predicted. The equation works if the investors are sophisticated and well informed. It works if they're market newcomers just playing hunches. It even works if they're merely flipping coins.

"This is a true zero-intelligence model," says Farmer. "We're not concerned with people's rationality or their reasons for placing the trades they do. The equation works either way."

Farmer sees a similar hand of physics in the way even the stablest markets tend to drift from equilibrium toward some kind of periodic collapse—paralleling closely the way geologic and meteorologic systems drift toward the occasional earthquake or hurricane, and even paralleling the frequency with which the turbulence happens. "It's as if something's pumping energy into all these systems," he says. He even sees something curiously Newtonian in the way markets in motion tend to remain in motion for a while—even if circumstances change that should slow them down—and markets at rest tend to remain at rest. Says LeBaron: "The study of physics has been moving into the study of markets for a while now. It's hard to tell what their boundaries are."

BUT ARE THINGS really as insensible as all that? There's certainly an appeal to such a scaling down of market science, if only because it makes it more comprehensible and creates the illusion that it can be made more controllable too. However, we may be stuck with more complexity than we'd like. For all the hard equations that appear to govern monetary dynamics, even the best modern-day economists still have to resort to a lot of trust-me hedging when explaining why economic systems work the way they do. Adam Smith posited his "invisible hand"—the leveling force that stabilizes markets—in 1776, and today's economic theorists still invoke it. Complexity scientists studying the markets may sometimes rely on similarly vague ideas, merely borrowing laws of physics and applying them to finance more as analogy than fact. "Some of these things are just metaphors," admits LeBaron. "We really don't know why they work or when they work."

One of the reasons they work is that invisible hands are still governed by invisible brains—undeniably complex ones. Capitalism functions reasonably well and communism flopped so badly because one system includes human needs, motivations, and occasional wisdom in its equations and the other stripped them out. Many complexity theorists studying stock trades argue that collective human brainpower is an underestimated thing, and the more minds you put to work in a market, the better you're all likely to do.

Brooke Harrington, a sociologist and professor of public policy at Brown University, believes that if you really want to understand the extraordinary potential of combined brainpower, you need look no further than the jelly bean contest. There's a reason trying to guess the number of jelly beans in a jar is such a venerable game, and that's because it's such a hard one. We do a great job of estimating things when they're on a scale that's familiar to us. Go to enough baseball games and you can get pretty good at judging what a crowd of fifty thousand people looks like. But fifty thousand bees? The same goes for, say, intuitively knowing the difference between someone who weighs 120 pounds and someone who weighs 150. Try to make the same judgment about the weight of two elephants, however, and your miss-distance might be on the order of tons.

Jelly beans are equally confounding, and when we try to guess how many are in a mass of them, we're almost always way off the mark. But when a lot of people guess, it's another matter entirely. Inevitably, Harrington says, the average of everyone's guess in a jelly bean contest will come significantly closer than any one person's guess. It doesn't merely happen sometimes; it doesn't happen even most of the time. "It happens virtually every time," Harrington says. "It's freakish, and it still amazes me."

Harrington stresses that more than simple laws of averages are involved here. After all, everybody could just as easily guess low or high, meaning that the group average would then be farther from the true figure than the handful of players who guessed closer to it. Instead, we all guess in such a precise distribution around the target that together we practically hit it. More important, the greater the number of people participating, the closer the collaborative guess becomes. Harrington does not pretend to know precisely the mechanism that drives such precision, but there's clearly something more complex than a vague invisible hand. "It's the old story of the wisdom of crowds," she says.

Harrington wanted to pursue this more-is-more finding further and decided to see if it applied equally well when people gather into investment clubs—increasingly popular groups that bring largely inexperienced traders together in the hope that the stocks they pick collaboratively will be better than the ones they would pick alone. Over the course of a year, she conducted a pilot survey of seven investment clubs in the San Francisco Bay area, then expanded it to a more comprehensive nationwide study of 1,249 clubs comprising more than eleven thousand members. What she discovered confirmed precisely what she suspected she'd discover: In general, the average group did better than the average individual; and larger, more diverse groups did better than smaller, more homogenous ones. Over and over again, a ragtag band of amateurs could do a better job than even many individual professionals in determining where the greatest number of investment jelly beans were to be found. "Trying to invest in the market when there are seven thousand available stocks is a formidable choice space to negotiate," Harrington says. "The more complex the problem and the more complex the environment, the more diverse points of view you need."

Lay traders are shrewder than they get credit for in other ways too. One of the great bits of stock market wisdom is that investors of all kinds are a jumpy breed, easily frightened by bad headlines from the world outside Wall Street. Give them a sudden bulletin about a natural disaster or an assassinated leader and they'll run amok, cashing out of the market in the self-fulfilling belief that everyone else is about to do the same. This don't-spook-the horses dictum arises straight from the herding and schooling theories, and while the dark experiences of 1929 and 1987 prove that it does apply sometimes, most of the time it simply doesn't.

In 1989, a team of three economists that included then-professor Lawrence Summers—well before his turbulent tenure as president of Harvard University—selected the forty-nine most newsmaking events from 1941 to 1987, including the attack on Pearl Harbor, the assassination of President Kennedy, the attempted assassination of President Reagan, and the Chernobyl disaster. Their purpose was to determine if those news jolts indeed shook the markets, and if so, how much. The researchers used the Standard & Poor's Index as their yardstick of market response, since this indicator seemed particularly sensitive to external shocks. What they found, however, was that even on the S&P, bad news in the papers did not have to mean bad news on Wall Street. The biggest one-day loss among all of the forty-nine days they surveyed was a 6.62 percent drop on one trading day in 1955 when President Eisenhower suffered his heart attack. This was little more than a flesh wound compared to the 23 percent hemorrhage of the 1929 crash and the nearly equivalent losses in 1987. The next biggest drops were a 5.38 percent hit when North Korea invaded the south in June 1950 and a comparatively minor 4.37 percent loss following Pearl Harbor.

Beyond those setbacks, Wall Street took news punches extraordinarily well. The Kennedy assassination led to a 2.81 percent loss. Four days later, after the transition to President Johnson proved smooth, the market actually rose 3.98 percent. The Reagan shooting? A quarter-point loss in a short day of trading and a 1.25 percent gain the next day. Chernobyl? A one-point loss. The Soviet invasion of Afghanistan? The market actually gained a point.

The study next looked at the data the other way, selecting the fifty biggest one-day movements in prices from 1946 to 1987 and pairing those changes up with what was in the papers those days. In only nine of the cases did a nonfinancial news story like international tensions or a political shake-up seem to account for the market movement. In all of the others, prices appeared to be moving in response to such dry-as-toast news as a fall in short-term interest rates, the end of a stubborn coal strike, and a rumor that President Nixon was taking steps to reenergize his then-stalled economic policies. In one case in June 1962, *The New York Times* accounted for a 3.5 percent jump in stocks with the simple horse-sense explanation that "Stock prices advanced chiefly because they had gone down so long and so far, a rally was due." In all of these cases, it wasn't physics or mathematical formulae that were keeping the markets from listing, but the far more complex variables of human intelligence and levelheadedness.

There is, of course, one other exceedingly complex and uniquely emotional variable individual traders bring to the market equation, one that almost entirely defies parsing: a sense of fair play. Even the most stripped-down market models take on many more dimensions when you factor in the human nose for what constitutes a square deal and the sense of outrage that arises when it looks like the bad guys are going to get away with something. Mutual fund managers may be only too happy to pick up tobacco or oil company stocks when the market looks good for them, but they are also smart enough to keep a close eye on them, knowing that when the periodic waves of bad press hit, traders recoil from buying the stocks, as much because of the companies' possibly dwindling value as because of the foul taste a share of Philip Morris or ExxonMobil suddenly leaves. Investors who cleaned up during the long ascent of Enron might not be willing to divest themselves of the revenues they made now that the company has crashed and burned, but given the opportunity to do the whole thing over again, it's a fair bet that a good number of them would put their money elsewhere, unable to choke down profits that they know were earned at the expense of elderly Californians who went without power or Enron employees who were fleeced of their savings.

Researchers have long wondered just how powerful the human fairness impulse is, and while it may never be truly quantifiable, some studies have shed some light on it. Samuel Bowles, a professor of economics at the University of Siena in Italy and a faculty member at SFI, has conducted a simple fairness test in numerous societies around the world and has been surprised by both the findings themselves and their dependable consistency. He begins the experiment by giving two subjects a quantity of money to share. Sometimes it's a relative token amount like twenty dollars; sometimes it's a more substantial hundred dollars. In a number of cases it's a truly meaningful figure like three months' salary. In all cases, the money is real and the subjects do keep it at the end of the experiment—provided they can agree on how to split it up.

Bowles allows the first of the two subjects to choose how the windfall will be divided—fifty-fifty or something less equitable. The second player then decides either to accept the cut—in which case the experiment ends and the money is paid out accordingly—or reject it, in which case the experiment also ends but nobody gets anything. Economists, for whom the idea of the selfish economic man has always been a central theoretical pillar, would predict that the second player would accept almost any division of the pot, since a little free money is better than none at all, even if the other guy clearly got the better of the deal. That, however, is not what happens.

Over and over and over again, Bowles found that anything that falls too far short of an even split is rejected outright. The average accepted bid is 43 percent. Offers of 25 percent or less are almost never accepted, even if the amount is nonetheless considerable. Bowles has run his study in seventeen countries and taken pains to include populations entirely unfamiliar with the idea of organized modern markets, such as the Ache hunters of eastern Paraguay and the Machiguenga Indians, hunter gatherers in the Peruvian Amazon. In those cultures too, the reaction to a bum deal was the same as in commercial societies. What's more, when the brains of players are scanned with magnetic resonance imaging machines during the experiments, the pictures that result reveal that the neural regions activated by an unfair offer are the same ones activated by feelings of disgust. That's not just a dispassionate reaction to an inequitable bargain, that's a primal recoiling.

"People turn down a lot of money," says Bowles. "When you ask them why they reject a 20 percent share, they answer, 'Because that son of a bitch was getting 80 percent.' So people are willing to pay 20 percent just to punish a son of a bitch." While Bowles's studies explore only what happens when the participants in the game themselves are the ones getting shortchanged, the almost universal schadenfreude we feel when a white collar criminal is perp-walked off to trial suggests that our sense of principled outrage doesn't stop at our own bank accounts. "We do appear to be wired for justice," says Bowles.

Such a sense of fairness is not a thing calculable in a physics equation; neither are the instincts at play in a jelly bean contest, nor are the collaborative hunches that operate when people come together in investment clubs. That is where, finally, parsing the market as either simple or complex becomes such a hard theoretical nut to crack. Markets are capable of moving toward greater and greater order, synthesizing new wealth out of the mere act of exchanging existing wealth, the same way that plants manufacture new matter—leaves, stems, flowers—by stirring together existing elements they absorb from water and carbon dioxide and activating them with the energy of the sun. If that's not complex, nothing is. And yet, plants die and markets crash, leaves dry up and wealth vanishes—both examples of order collapsing into disorder and the complex withering down to the simple.

In the case of the markets, it's the interplay of human minds that mostly determines which way things will go, and this is where the business of being confused by everyone else can become so tricky—and where there are lessons that apply far outside the markets. The mad rush of dozens of states to move their presidential primaries earlier into the election cycle is a perfect example of mass thinking gone mad. The states' perfectly reasonable objective, of course, is to share some of the outsized influence wielded by the Iowa caucuses and New Hampshire primary, which every four years establish the dominant candidates in both parties and eliminate some of the laggards before the other forty-eight states can cast so much as a single vote. The problem is, every state that joins the newly front-loaded scrum dilutes the value of voting so early. This increases the possibility of a deadlock, with each candidate picking up a victory here or there and no one collecting enough to emerge from the pack. The paradoxical result could be that those states that decided to wait—ostensibly dooming themselves to electoral irrelevance—will wind up breaking the logjam and deciding the matter.

The same kind of pattern plays out in real estate migrations, in which people flee cities for suburbs seeking more space and better value, only to see sprawl increase, traffic explode, and prices soar, as the new emigrés simply overload the system. Meantime, prices fall in the abandoned city and those people left behind snap up homes at a fraction of the prices they went for before. Fashion, music, and movie trends follow similar boom and bust patterns. Knowing when to ride the wave of hip-hop, action films, and seventies revival clothing is as important as knowing when the market is saturated and it's time to move onto something else. And remember when American car buyers wanted nothing more than ever-bigger SUVs? How'd you like to be a dealer stuck with a lotful of them now?

The key in all of these situations is to recognize that there is indeed great wisdom in what everyone else knows. But there is sometimes greater wisdom in knowing what only you know. There may be no such thing as mastering the perfect balance of those skills; but there is such a thing as becoming powerfully—and, in the case of the markets, profitably—good at it.

Of course, there are times when more than your portfolio is on the line, when more than stocks or products or even presidential elections are in play. There are times when your very life is the commodity at risk. Learn all you want about following or not following the crowd, but what happens if the rules seem to vanish at once, if you don't know how to move and nobody around you seems to know either? The rules of simplicity and complexity may make you rich, but the next step is knowing when you have to use them to keep yourself alive.

CHAPTER TWO

Why is it so hard to leave a burning building or an endangered city?

Confused by Instincts

ED SCHMITT WAS NOT IN THE HABIT of ordering up a Jim Beam at ten o'clock on a weekday morning—and he certainly wasn't in the habit of ordering a second one at ten-fifteen. When he did drink, he at least had the sense to finish up before leaving the bar, rather than carrying his glass with him out to the street. But this morning he did just that, stepping straight out of Nathan Hale's saloon on Murray Street, drink in hand, taking care to cover the glass lest his bourbon be ruined by the grime raining out of the sky—grime that was all that remained of the building where he had been settling in at work less than two hours before.

Until this morning, Schmitt had been a reorganization clerk at the Fiduciary Trust Company on the ninety-seventh floor of the World Trade Center, Tower Two, in New York City. His day had begun as all of his workdays began, with a bacon, egg, and cheese sandwich and a cup of coffee from Akbar's Greenhouse Café in the basement of the tower. Schmitt made the long elevator ride up to his office and collected what he and his coworkers called the Truck, a lockbox on wheels that contained the company's stock certificates. Rolling it toward his desk, he passed down an aisle that fronted on the building's north windows, and there he saw something grotesque: a mammoth passenger jet flying absurdly low, speeding straight toward him down the midline of Manhattan. This was all the wrong altitude for such a plane and all the wrong airspace in any event—the corridor over the Hudson River being the proper place for most air traffic.

Schmitt stared at the approaching thing and did . . . precisely nothing, standing at the window, holding onto the handle of the Truck, coolly reflecting that he was almost certainly about to die, and curiously accepting that uncomfortable fact. But at the last moment, the plane slewed to the side, banking at a sickening angle and heading straight for the flank of Tower One, into which it vanished. A fireball appeared in its place.

The violence of the explosion made Schmitt jump, but oddly did not stir him from his window. What did move him was the sound of Doris Torres, a coworker, standing up at her desk just a few steps from where he stood. He turned to her, saw her collecting her things, and nodded in agreement: "I'm getting out of here," he said.

Torres and Schmitt both moved for the exits, but they noticed that a lot of other people didn't. Everyone in the office had leapt toward the windows at the sound of the impact, but after that, there was a lot of milling, a lot of muttering. Some people turned on the radio or TV; some picked up the phone. A vice president on the south side of the floor persuaded his secretary and other members of his staff to stay at their posts. Schmitt and many others filed into the unexpectedly narrow stairwell and began making their way slowly down.

Progress on the crowded stairs was incremental. People joining the flow at lower floors slowed things down from their already creeping pace. So did older people; so did obese people—their gauge too wide for the two-abreast maximum of most of the stairs. Debates echoed up and down the stairs about whether it was wise to continue on, if it was worth braving the bedlam outside. Some people bailed out to hunker down on whatever floor they happened to be passing when their nerve failed. Schmitt pressed on, and when he had descended to the sixty-somethingth floor, an announcement came through the stairwell speakers, instructing the Tower Two workers to stay at their desks. "The situation is confined to Tower One," the voice reassured. Still more people abandoned the stairwell. When Schmitt reached the forty-third floor, a terrific concussion shook Tower Two and a muffled roar echoed up and down the stairwell. People began pushing. "Relax," Schmitt said to someone next to him, "that was probably an electrical explosion caused by the fire next door." He had no way of knowing if it was anything of the kind.

When Schmitt and the others spilled out of the stairway into the building's second-floor lobby, they were herded between flanks of Port Authority police in a sort of human cattle chute, under the Trade Center plaza and up onto the street about fifty yards away. Freed into the light, the evacuees looked up at the holes burning in the towers and scattered in all directions, Schmitt to Nathan Hale's, where he had his Jim Beam, ordered his second one, watched on the bar's TV as the two towers fell, and cursed his folly as the owner of the saloon slammed the door against the cloud that seemed certain to bring the little building down in the shock wave of the big ones. When the densest part of the dust storm subsided, Schmitt finally left, eventually tossing his fool drink aside in the ruined streets, and walked up to midtown, where he at last felt safe.

Ninety-seven of Ed Schmitt's coworkers died that morning, including the vice president who'd cajoled his staff into staying. Doris Torres, whose simple act of standing at her desk had stirred Schmitt from his daze at the window, had chosen to evacuate via elevator instead of stairwell, but courteously stepped aside to allow others to board what turned out to be the last car that left the ninety-seventh floor. She got a late start descending by stair and made it to the street alive, but sustained burns on the way that claimed her less than a week later.

The people who stayed behind in both towers on September 11, 2001—or waited too long before trying to leave—bore no responsibility for what happened to them that morning. They were, instead, twice vicitimized—once by the men who hijacked the planes and took so many lives; and once by the impossibly complex interplay of luck, guesswork, psychology, architecture, and more that is at play in any such mass movement of people. Fear plays a role, so does bravado, so does desperation. But so do ergonomics, fluid dynamics, engineering, even physics—all combining to determine which individuals get where they're going, which ones don't, and which survive the journey at all. Ultimately, we're misled by our most basic instincts—the belief that we know where the danger is and how best to respond to keep ourselves alive, when in fact we sometimes have no idea at all. It's the job of the people who think about such matters to tease all these things apart and put them back together in buildings and vehicles that keep their occupants alive. It's the job of those occupants to learn enough about the systems so that they have the sense to use them.

TO SCIENTISTS STUDYING complex systems, evacuees abandoning a building or city or disabled airplane are not so much humans engaged in mortal flight as data points on the complexity arc. On an ordinary day, twenty thousand people working in a skyscraper or half a million people in a coastal city occupy the same spot on the complexity spectrum as air molecules filling a room—moving randomly and chaotically in all directions, filling all the available space more or less uniformly. They're very active, but also very simple and disordered. Send the same people on a stampede down stairways or onto highways and things quickly grow overloaded and grind to a halt, jumping to the other end of the complexity arc—robust, unchanging, frozen in place, but every bit as simple as the ever-shifting air molecules. It's in the middle of the arc, where the molecules just begin to take some shape or the people in the tower just begin to move to the exits, that true complexity begins to emerge.

The best way to understand the elaborate manner in which people move en masse may be to understand the equally complex way water does the same, particularly how it navigates around obstacles or breakwaters. A foundered boat or a tumbled boulder in the middle of a rushing river turns even the most powerful current chaotic, reallocating its energy into increased swirling and churning and decreased velocity. When people are fleeing a building, a similar kind of chaos can be a very good thing. Designers of interior spaces have found that a perfectly useless post positioned along the path to a fire exit may actually help people escape. Give evacuees storming toward a doorway a little something to avoid and you stagger their arrival slightly, allowing them to stream through the opening in a reasonably controlled flow, rather than colliding there at once and causing a pileup. The obstacle keeps you at the top of the complexity arc, preventing you from plunging headlong to the frozen end.

"You create a little turbulence," says Santa Fe Institute economist John Miller, who specializes in complex adaptive social systems. "By adding a little noise to the system you produce coherence in the flow."

The problem with the human-beings-as-water-particles idea is that it takes you only so far. Early fluid-based computer models that designers relied on to simulate crowd flow had a troubling habit of producing results that were simply too tidy. Put a single exit in an office and the computerized evacuees would flow through it at a particular rate of speed. Open a second exit somewhere else and the people would respond appropriately, with half of them choosing that new option and the flow at both doors adjusting itself commensurately. Put a breakwater somewhere along the route and things smoothed out even more. Certainly, the programmers weren't fools. They would correct for things like accessibility and proximity of the exits—with more of the simulated evacuees choosing the closer, more convenient egresses. But after that, the software assumed, the people would behave sensibly. This, of course, was nonsense.

For one thing, people have different levels of decision-making skills, with some inevitably behaving more rationally than others. For another, all of us have a tendency to believe that the rest of the group knows what it's doing, and thus will gravitate toward a more popular exit simply because other people have chosen it, even if the alternative is perfectly safe and much less congested. Every additional person who chooses the more popular exit only makes it likelier that the next person will select that one too. Finally, information tends to get distributed unevenly, with some people learning about an emergency first and acting before the others. Once all of this was written into the fluid-based programs, the simulations ran amok. "A fluid particle cannot experience fear or pain," writes David Low, a civil engineer who studies evacuations at Hariot-Watt University in Edinburgh, Scotland. "It cannot have a preferred direction of motion, cannot make decisions, cannot stumble or fall." Humans do all of those things and, in the process, make a mess of most models.

In response to such problems, evacuation software has gotten a lot more sophisticated. Programs with such evocative names as EGRESS, EESCAPE, and EXODUS take into consideration everything from the nature of an emergency (fire, bomb scare, blackout) to the season in which it occurs (cold weather causing more problems since people have to collect their coats, which is not only time-consuming but space-consuming once the bulkily clad occupants of the building crowd into the stairwells) to the time of day (fewer people being in the building at lunchtime, for example, than at ten-thirty in the morning, meaning that the midday evacuations will be quicker and smoother).

Of course, computers can't model the existential terror that comes along with any evacuation and programs like EGRESS are thus limited by their very nature. One place those limitations show themselves is in the evacuation stairway. Once frightened people reach the stairways that should take them to safety, things ought to get easier. To complexity scientists, emergency stairways are nothing more than what are known as relaxation pathways, outlets that allow pressure or an imbalance to be relieved—in this case, the pressure of a mass of people trying to descend from higher floors to lower floors and then spill out of the building altogether. Given that, the only thing that ought to count in a stairway evacuation is pure speed; anything short of a stampede that will get the building emptied in a hurry is a good thing. But pure speed is hard to maintain.

Much has been made in the years since September 11 about the announcement in Tower Two that instructed workers to stay at their desks—something that undoubtedly contributed to some of the deaths that day even though that was precisely the correct thing to do as far as anyone at the moment knew. The improbability of one building suffering the kind of violence that was done to the towers that day—to say nothing of two—called not for a wholesale evacuation, but for a so-called sheltering in place, taking cover precisely where you are and thus staying safe from the chaos and debris on the street below. What's more, ordinary fires don't spread instantly across multiple floors and aren't fed by thousands of gallons of jet fuel. Fireproof doors and localized sprinklers are designed to keep people on unaffected floors safe so that they don't crowd the stairwells and slow the people who are in real danger.

Even in an unexceptional emergency, however, it doesn't take much for a smooth stream of downward-flowing evacuees to turn turbulent. For one thing, people from middle floors who enter the stairwell somewhere in mid-current can cause things to slow or stop. The delay flows back along the queue the same way a ripple propagates through water, or cars entering a highway briefly slow the ones closest to the on-ramp, causing a wave of tapped brake lights that may radiate miles back. Sequential evacuation is the best way to handle this problem, with people on lower floors leaving the building first and each higher floor following successively. But again, nobody pretends that people fearing for their lives would wait patiently at their desks until their floor was called, nor would it be wise in a situation like the September 11 attacks, when the towers themselves turned into ticking clocks, and people higher up needed every minute of evacuation time they could get before the buildings collapsed on top of them. Instead, it's the design of the stairways themselves that must provide the extra margin of speed.

After the 1993 truck bomb attack on the World Trade Center, New York City officials conducted an exhaustive study of what went right and what went wrong in the evacuation that followed the blast and how to improve things in the future. One of the first things the investigators decided was that the stairways' handrails needed fixing. Most handrails are designed with sighted people in mind and are thus intended to provide physical support and nothing more. As a result, they often end just a few inches before each landing is reached, since there's little need to hold yourself up when you reach the bottom of a flight. But that doesn't work as well for sightless people, who rely on the tactile cue of the handrail to know precisely how many steps they must negotiate. And it works equally poorly when sighted people are groping blindly in a crowded stairway in which the lights may be flickering on and off. In those situations, a handrail that quits one step short of the bottom is a very bad thing. Survivors of the 1993 bombings widely complained of the pileups at landings when evacuees would stumble over a last stair that the rails suggested wasn't there at all. In response to those reports, the handrails were extended.

Lighting was improved too, thanks to more powerful illumination and better generators kicking in if the main power source failed. And fluorescent tape was applied to all of the stairs, with strips showing where each step stopped and arrows pointing the way around the landings to the stairs leading farther down. The tape, in fact, may have been overapplied, with far more used than was strictly necessary. But the designers were responding not just to the evacuees' navigational needs, but to their more complex emotional ones. Light, any light at all, has a calming effect on frightened people, all the more so when it's applied in a reassuring shape that points the way out of the danger.

"The tape was a yellowish-orange," says Schmitt, who followed ninety-seven floors' worth of arrows down to safety. "Looking down and seeing it, even when the lights were on, was one of the most helpful things."

One thing the 1993 investigators couldn't change (and the towers' 1960s designers couldn't foresee) was how inadequate the width of the buildings' stairwells would be, more so in 2001 than ever before. Two of the three stairwells in each of the World Trade towers were just forty-four inches wide; the third was a somewhat more commodious fifty-six inches. They were designed in part based on what Jake Pauls, a Baltimore-based engineer who heads a building safety consulting firm, calls the "flawed traditional assumption" that people running for their lives will walk two-by-two from the top floor of a skyscraper down to the bottom, making maximum practical use of the stairway space that's given them. This was considered illusory by many people even forty years ago, a point that some safety advocates made to the buildings' designers before the towers went up. The architects yielded a bit, which is how the buildings each wound up with one fifty-six-inch stairway, but they held the line on the other two. What the safety advocates failed to reckon on in the 1960s was that come 2001, more than two-thirds of Americans would be overweight or obese, meaning that they would not only take up more room in already narrow stairwells, but that even without crowding, they'd be struggling more to make the long climb down.

"The number of people who have trouble with stairs these days has increased by a factor of between two and six," says Pauls. "Studies conducted after September 11 showed that the performance of people coming down the stairs was twice as slow as when we did similar studies in the 1960s. And those studies were themselves two times slower than the assumption upon which the design of the stairs was based. The result now is that evacuation is only a quarter as fast as what we thought when the forty-four-inch stairway was built."

The answer, Pauls believes, is universal stair widths of at least fifty-six inches and ideally ten inches more than that. And while it would be better all around if Americans slimmed down, the business of saving lives requires creating buildings for occupants as they are, not as they presumably ought to be.

DESIGNING AROUND THE physical constraints of body and building is actually the simpler part of the evacuation equation. Far more complex are the social parts. Most social systems operate under a very helpful set of assumptions that the actors within them quickly learn and obey. In an office, it's the highest ranking person who makes the rules. In a retail space like a department store, it's often one of the lowest-ranking ones— the one in a security guard's uniform. Powerful people flying first-class readily subordinate their will to the complete, if temporary, authority of a flight attendant. Business executives in a driver's license queue obey the orders of city-wage clerks. Other assumptions, beyond questions of heirarchy, are at work too. Residents of heavily populated places like Tokyo or New York, for example, often have very rigid rules about personal space, agreeing to stay a fixed distance apart to preserve what breathing room they can in their beehive cities. Even in an emergency, the no-touching habit can be hard to break, often slowing things down as people reacclimate themselves to a suddenly new protocol.

But broken these and other norms do get. What appears in their place in a time of emergency is something complexity researchers call the emergent norm—that is, a whole new set of rules arising from the circumstances. The first thing people often do in a relative slow-motion emergency like a blackout or a fire is begin what appears to be the manifestly wasteful practice of milling, standing about and quizzing one another about a problem none of them really understands. If televisions, radios, and phones are available, people will use them to gather more data still. This is what went on throughout the twin towers on September 11, and while in restrospect it seems that all of the residents should have fled the instant after the first impact, when the events are unfolding, things aren't nearly so clear.

"Milling is sometimes seen as a bad thing," says sociologist Benigno Aguirre of the University of Delaware's Disaster Research Center. "But people aren't robots or herding animals. Even given difficult situations, they continue to absorb information and act on what they learn."

Complexity researchers call the introduction of ideas in this ad hoc system "keynoting," tossing out a thought around which the discussion crystallizes. Schmitt's simple declaration, "I'm getting out of here," was a keynote. So too was his mistaken guess that the sound of the second plane striking Tower Two was just an explosion from next door. This idea was believed by people only too happy to accept so benign an explanation, leading to a new action—or constructive inaction—with most of them continuing down the stairway, avoiding panic that could have doomed them all. "Somebody comes up with a keynote," Aguirre says. "Counter-ideas or theories come forth. Trial and error happens. And eventually an emergent norm is reached."

Like so many other processes in the complexity field, this one has a handy way of mirroring other, seemingly unrelated phenomena. In this case, Miller says, the emerging norm is strikingly similar to the process of annealing metals. In their natural state, molecules of metal are arranged chaotically. Melt the mass down, and this disordered state orders itself a bit. When the material cools, the molecules jumble up again—but a little less so than before. Perform the heating-and-cooling cycle just right, and the molecules reorganize their ranks, settling into something closer to order. The process essentially bakes the brittleness out of the metal but preserves the flexibility, which is why metallurgists go to all the trouble in the first place. A strong but flexible consensus is precisely what a group of people in crisis is looking for too, and it's precisely what keynoting and milling achieve. "The atoms in the metal want to align with one another just like the people in the group," says Miller. "By introducing noise into the system and then cooling it back out, you let that happen."

Of course, some metals—and some groups—anneal better than others. People in office towers reach agreement a whole lot faster than people in high-rise apartments. One reason is that you're much less likely to be undressed in your office than you are in your home, where you may be coming straight from the shower or bed at the instant an emergency hits. When you're less ready to evacuate, you're also less inclined—and far likelier to convince yourself that the fire or the blackout isn't really so bad. For another thing, you're much less invested in the objects in your office. It's a lot easier to abandon ship when all you're leaving behind is a cabinet full of files and a company-owned computer. That's not the case when you're saying good-bye to your wedding album, family photos, and grandparents' heirlooms.

Behavioral differences even emerge between people in various types of apartments. A 1996 study by the National Research Center in Canada showed that the taller a building is, the less milling and annealing take place during an emergency, mostly because there are simply more strangers in a high-rise structure than there are in a smaller, more intimate one. This can lead to bad evacuation decisions as isolated individuals act on limited knowledge, darting down the closest stairway, for example, when just a few questions would have revealed that one was choked by debris or smoke. On the other hand, once people do get moving, they tend to go slower in smaller buildings, with the debate and discussion often continuing past the point of usefulness—nudging it over the top of the complexity arc, where annealing leads to congealing.

Many of the quirks of human behavior at play when people are fleeing a burning building are in even greater evidence when they're abandoning a disabled airplane, where the evacuees are packed much more closely and death can come much, much more quickly. For all the headlines airline evacuations make, they're actually quite common, with an average of eleven taking place in the United States every day, most for problems as comparatively manageable as a balky tire or a wisp of smoke in the cabin. The reason you hear about so few of them is that even the more serious ones usually end uneventfully.

"Most airline incidents are extremely survivable," says Cynthia Corbett, an experimental psychologist and human factors specialist for the Federal Aviation Administration. "On average, I see three or four a day, but all most people hear about are the serious or fatal ones."

But if airline emergencies don't usually claim any lives, it's often no thanks to the passengers themselves, who show a mulish tendency to behave in ways most likely to kill them. FAA investigators regularly run simulated evacuations at the agency's training center in Oklahoma City, trying to study the panicky scrambling that occurs in emergencies. The job is not easy, since even with a dummy airplane cabin and the best faux smoke and other special effects, the subjects know that no one is really in any danger and behave with a decorum that simply doesn't occur in real life. To elicit the disorderly behavior that so often breaks out during the real thing, the investigators play not on the subjects' fears, but on their greed.

Volunteers in FAA studies earn about $11 per hour for their efforts. Run a simulated evacuation in which the first 25 percent who make it off the plane receive double that amount—still a pretty modest payout—and suddenly people are getting jostled and elbowed as the volunteers push to qualify for the bonus. In the United Kingdom, similar experiments are run, but winning passengers don't have to wait for their checks as they do in the United States. Instead they are handed five-pound notes the moment they reach the tarmac. Anticipating an immediate payout, they generally behave even worse. In both cases, it's not the financial bounty itself that motivates people so much as it is the mere fact of the competition and the experience of receiving a reward, no matter how modest. But whatever the cause, the tests produce something much closer to the results the investigators need—less urgent than a real emergency, perhaps, but still very unmannerly.

"The British investigators videotaped one of the simulated escapes and showed it to survivors of a 1985 accident in Manchester in which many passengers perished trying to struggle through the over-wing exits," says Corbett. "The survivors said that's exactly how the people on their plane behaved."

One thing that can make a difference, at least in terms of reducing the panic, is for the passengers to tend to their own safety before they leave the ground. It may be true that not all airplane accidents are the death sentences people think they are, but about half of the passengers in all true emergencies or crashes do die. Investigators are convinced that the people who make it out alive are often those who had an evacuation plan in mind from the start, particularly looking fore and aft to determine where the closest and second closest emergency exits are and keeping that information in mind throughout the flight. The few seconds you don't spend searching around in the dark as the cabin fills with smoke or water are the few seconds that could save your life. Similarly, for all the yawns the flight attendants' familiar safety lecture elicits, the fact is most people have never really absorbed what they've been told. Few airline passengers really need any help knowing how to work a seat belt, but the operation of the flotation vest and oxygen mask is still a mystery to most fliers.

"Not knowing what to do can lead to freezing behavior," Corbett says—just the kind of dazed inertia that kept Schmitt rooted at his Tower Two window when Tower One was attacked. "There's a reason business travelers, who fly all the time, have a better survival rate than occasional travelers." And how to get people to pay attention for once when the flight attendants are speaking? Corbett's at a loss. "Maybe send them to websites and give them free miles if they can pass a safety test," she suggests.

IF HUMAN BEINGS can be vexing things when they're passively riding aboard an airplane or occupying an office building, how much more complicated do they become when you give them vehicles of their own and set them loose to operate them at will? That's a question that gets tested, if never quite answered, on the world's highways every day. If ever there was a laboratory for the pinball interplay of physics, psychology, and overall complexity, it's in the chaotic scrum of daily traffic.

Few people understand the elaborate dance of traffic better than Sam Schwartz, known to the builders and city planners who regularly retain the services of his sixty-person consulting firm as Gridlock Sam. Gridlock Sam got his name in 1980 when, as a young urban planner working for New York mayor Ed Koch, he was helping the city prepare for what turned out to be an eleven-day transit strike. Writing up a memo of emergency recommendations for senior officials, he recalled the words of a colleague several years earlier who had been analyzing a proposal to close Broadway to vehicular traffic. His colleague gave the plan the thumbs-down, worrying that it would simply "lock up the grid." Schwartz was always struck by that image and titled his 1980 memo "Gridlock Prevention Plan." While New York officials could take or leave some of his ideas, New York commentators loved the newly minted word, pronouncing it crisp, visual, a privilege to pronounce—and credited Schwartz with its creation.

"I wasn't even the first person to put the two images together in a sentence," he says today. "All I did was put them together in a word."

No matter, it's his now, as is the mantle of the world's leading authority on traffic. One thing Schwartz has learned over his decades in the field is that when it comes to traffic, physics comes before everything else. And once again, it's the complexity arc that best captures things.

When traffic on a highway is very light, cars behave like the air molecules in the room, doing whatever they like, going wherever they please. When traffic is very heavy, they collect into something closer to the molecules in the frozen carbon—going almost nowhere at all. It's chaos at one extreme and unchanging robustness—like the chilled carbon—at the other. Neither one is very complex. But things change at the top of the arc, up in the area where average speed ranges between twenty-five and forty-five miles per hour and the volume of vehicles ranges between 5,000 and 6,200 cars, buses, and trucks passing a given point in that same hour. At this spot in the arc, things are precariously balanced, with almost anything able to tip them one way or the other. A single driver in a single car reaching for a coffee cup and absently tapping his brakes can trigger a ripple of taillights that instantly turns into a clot of slowed traffic. A few people exiting the highway can open a clear spot that relieves pressure and increases speeds for miles.

"A favorite cartoon of mine shows a congested highway and a sign that reads, 'Traffic inexplicably clears ahead,'" says Schwartz. "But in fact it's very explicable. You have this zone of instability in which a single change can cause a new speed to gel instantly."

One way to thin things out, easing them toward the less dense end of the arc, is to control the way new cars enter the already swollen stream. Increasingly common are entrance ramps equipped with stoplights that can be switched on when traffic gets heavy and timed to change from red to green at different intervals depending on how bad that congestion is. Just like a river can hold an unlimited quantity of water, but only if it flows at the proper rate, so too can a highway handle any number of cars if they're fed through slowly enough. "Essentially," says Schwartz, "you slow things down to speed them up. It works on the same principle as fluid dynamics. Five cars entering a highway at once has far more effect on traffic than the same five cars, each arriving ten seconds apart."

Choke points get even narrower and complexity grows even richer when the traffic moves off the main arteries and onto surface streets, particularly when those streets are in the unforgiving channels of a city. New York—and, in particular, Manhattan—is perhaps the best of all places to study this phenomenon, not just because of the sheer volume of the vehicles that crowd its streets every day, but because its neat little grid plan makes things far easier to measure and model. Investigators who have done some of that basic work have been impressed by just how much time Manhattan spends at the peak of the complexity arc, with traffic moving reasonably well, but the tiniest change in cars arriving or leaving sending it sliding down one side or the other.

At first it would not seem that the city's traffic should be perched on such a fulcrum. More than 1 million vehicles stream into and out of Manhattan every day—enough for the city's streets to gag on, and often they do. But it's just a fraction of that 1 million that determines if things keep rolling. At any given moment on an ordinary afternoon, there are only about eight thousand cars in operation between Thirty-fourth and Fifty-ninth Streets in midtown Manhattan. Subtract just a few of those and traffic improves noticeably; add an equally small number and it slows to a stop. Indeed, in the worst circumstances, during the paralysis of a true gridlock, the streets would actually be 60 percent empty. All of the action—and all of the crowding—takes place in the intersections, with nothing at all getting downstream to midblock.

It's this argument—just how few additional cars it takes to send New York traffic into a state of cold-carbon inertia—that planners make year after year, when Manhattanites, fed up with battling one another for cabs, lobby City Hall to issue up to sixteen thousand more hack licenses. "You want the most miles per taxis, not the most taxis per mile," says Schwartz. "Add that many new vehicles and all you would do is provide seating in midtown Manhattan."

But traffic management isn't all physics and fluid dynamics; as with building evacuations, it's human behavior too. Physics, after all, doesn't account for the fact that in the week after daylight savings time goes into effect, accidents increase 7 percent, simply because people are driving on less sleep, or that parking tickets spike on the day after Thanksgiving, not because of increased shopping or driving, but merely because it feels like a holiday and drivers are less disciplined about feeding meters and reading parking signs.

Failure to account for such eccentricities can cause governments to waste tens of millions of dollars on infrastructure that does nothing to reduce traffic, and in fact only exacerbates it. This turns up all the time in the common urban planning phenomenon known as induced use. Dig another tunnel or build another bridge in an overcrowded city and at first the result is exactly what you hope for—spreading a fixed number of cars over more arteries and speeding things up for everybody. But faster driving encourages more people to bring their cars into the city, jamming things back to precisely where they were before. Similarly, eliminating roads increases congestion in the short run, but in the long run can actually reduce it, as drivers give up their cars and find other ways to get around the city. The 1970s plan to shut down Broadway to vehicles might well have locked up the grid, but only for a little while.

"Traffic is not like water, a set amount that will flow anywhere," says Paul White, an urban planner and executive director of the nonprofit group Transportation Alternatives. "You induce use by giving drivers too many options and reduce it by cutting some off. You don't even have to close streets. You could simply eliminate some lanes or convert them to pedestrian walkways. There is always an adjustment period of one to three months, which is why street fairs, for example, do increase congestion on surrounding streets, but this would eventually dissipate if the fair were open every day."

All this suggests that while human behavior may be impossible to mathemetize, it can at least be influenced—and that could save lives. Every year, about 1.3 million people worldwide die in automobile accidents and a whopping 50 million more are injured. The cost of such carnage—about half a trillion dollars annually—hits all nations hard, but especially those in the Third World, where traffic regulation is especially poor and the money to pay for all the injuries and deaths is especially scarce. The irony is, if the price tag for the accidents is high, the cost of preventing them in the first place can be remarkably low.

Much of the creative energy in road and urban design goes into what planners call traffic calming methods, a combination of physical, visual, and psychological strategies that both smooth the flow of the vehicles and soothe the tempers of the drivers. The first and most effective of these systems is the common speed bump—which designers have lately renamed the speed hump.

The human body is built tougher than it looks, often able to survive collisions of between fifteen and twenty miles per hour—which is the most nature ever estimated we'd need since that represents the average top speed at which a sprinting person could collide with something unmoving like a tree. But nature didn't bargain on cars. The majority of pedestrians hit by a vehicle moving at or below the twenty-mile-per-hour threshhold do survive the impact.

But as the speed creeps up, the survivability drops, until at thirty-five miles per hour, the death rate tops 80 percent. The question of living or dying can then turn on that extra fifteen miles per hour, and planners have devised numerous ways to knock that small amount off the speedometer.

In communities where traffic is poorly regulated, particularly in developing countries, residents routinely practice what's known as rogue traffic calming, placing obstructions such as barrels or sawhorses in the middle of roads, or changing the topography of the surface itself, digging potholes or building low berms. But precise speed bumps work the best. In a city with reasonably well-maintained roads, a speed bump will rise four inches above the surface, with slightly tapered four-foot-long approaches on either side. In the past, most cities did without the approaches, making the bump more jarring and the lesson more pointed for motorists. But speed is sometimes necessary for fire trucks and other emergency vehicles. The gradual grading accommodates them while still discourging speeding, and is what turned speed bumps to speed humps. A four-inch-high hump can usually get traffic down to about eighteen miles per hour. Add an inch and you take it all the way down to ten mph. On an especially dangerous stretch, it doesn't hurt to have two or three on a single block. However many you have, well-placed bumps can go a very long way. In Ghana, speed bumps installed in a single hot spot slashed overall accidents by 35 percent, fatalities by 55 percent, and serious injuries by 76 percent.

Placing an obstacle in the path of drivers isn't the only way to slow them down. Sometimes all you have to do is make them think there's one. Cities pinched for even the modest funds necessary to install speed humps are finding that you can achieve almost as much with a can of paint and a little deft artwork, creating three-dimensional markings that make a flat patch of road look like it's rising up—enough at least to cause uncertain drivers to apply the brakes. The problem with this on-a-budget strategy is that the lesson may not last. Motorists unfamiliar with the street will be fooled a couple of times, but after that, they'll speed right over. "Studies show that warning signs and other visual cues often work for only about two weeks before people start to ignore them," says Amy Pfeiffer, an urban planner working for Transportation Alternatives. "So there's some question as to whether the optical illusion would work much longer."

An alternative is to slip a little deeper into the human brain, playing on illusions we may not even be aware of. Bicycle lanes are often separated from other traffic by nothing more than painted lines, an imperfect barrier at best. Designers experimenting with different forms of striping have found that long vertical lines that follow the entire border of the lane are poorer at slowing drivers than a series of short, horizontal stripes, framing the lane like ladder rungs. These seem to do a better job of reminding drivers of the nearby hazard and inducing them to make the wiser choice of tapping the brakes a bit. "For some reason," Pfeiffer says, "the brain processes horizontal stripes as a better reminder to slow down."

Another way to sneak in under the radar of drivers' consciousness is to create environments in which they feel uncertain or confused. Urban planners in the Netherlands grew increasingly frustrated with the limited effectiveness of speed humps in residential areas, with drivers quickly speeding up again immediately after they passed the obstacle. Instead, the designers stripped off the hump and the asphalt as well, replacing them with brightly colored bricks and other unfamiliar paving. This created ambiguity as to whether the drivers were on a road, a pedestrian arcade, or even a play area, and that caused them to take better care—a very good thing, since in many areas, streets in fact do serve multiple functions, with people crossing and congregating carelessly and children using them as recreation spaces.

"Drivers simply didn't know how to behave," says urban economoist Walter Hook, of the nonprofit Institute for Transportation and Development Policy, "so they slow down." The risk of course is that pedestrians and children may be equally confused and indeed treat the newly paved area with even less caution than they otherwise would. This gives drivers even more reason to be cautious, but endangers people on foot.

Flat-out scaring drivers can work well too. Twenty years ago, New York was considering modifying the aged Williamsburg Bridge, in part because its narrow, nine-foot lanes were grossly below the modern safety standard of twelve feet. Schwartz and others, however, determined that widening the lanes could actually make the accident problem worse, since a confident driver is also a cocky driver, one more inclined to take risks. Better to keep everyone just a little spooked—and moving just a little slower.

All of the rules that operate in immediately life-threatening situations—whether it's the cataclysm of a terrorist attack or airline crash or the more retail risks encountered on the highway—operate everywhere else in our world as well. The little bit of fear that can keep you cautious on the Williamsburg Bridge is the same kind of thing you're trying to instill in a three-year-old when you deliberately tell a scary tale of what can happen if you wander away on the street or jump too rambunctiously on the bed. A child who is too frozen by fear to venture out onto the sidewalk or roughhouse around the house is not what you're trying to achieve—any more than you want an airline passenger in a disabled plane to stand terrified in the aisle. But just the right amount of fear can be both bracing and clarifying, keeping the slightly frightened person mindful of the risks and thus able to act more or less freely and more or less safely.

Milling and annealing are at work all the time too. What else is a board meeting or R&D conference than a group of people exchanging information and insights, keynoting ideas and tempering whatever action is eventually taken by exploring lots of options first? Sometimes the matter can get overthought—with ill-considered misgivings causing a good idea to be abandoned, in the same way a near and clear stairwell might be bypassed by building evacuees who heed an ill-informed warning that it's blocked by smoke. Sometimes there can be too few doubting voices, leading to a stampede toward a bad idea—whether a disastrous product or a deadly evacuation route. Most of the time, however, the opportunity for a variety of thoughts to be heard leads to a better, more orderly decision than any one group member could have made alone.

You can find the concept of induced use at work in the larger world too. If traffic on bridges and highways actually increases as you build more of them, so too may the long waits for a simple cup of coffee at a Starbucks lengthen, at least partly as a result of the fact that there seems to be one of the stores on every corner, drawing more and more people in all the time. And skyrocketing paper bills in most offices could be at least partly due to the fact that high-speed printers simply make it easier to create and revise documents, leading things in precisely the opposite direction from the paperless office that the computer revolution promised.

Even the physics of traffic calming has broader analogies and applications. The goal of the speed bump is not to stop traffic flow, but to tame and control it. The same kinds of processes are at work with stop-sell orders that tap the brakes on the sale of stocks and prevent markets from crashing; in dams and floodgates that permit rivers and canals to keep flowing but control how they do it; even in the simple interpersonal strategy of two people on the brink of a shouting match taking a breath or counting to ten, then resuming their quarrel with just enough heat and speed bled off that they actually may get something resolved.

THE CHALLENGE IN all these situations is to start with the already complex repertoire of human behavior, introduce it into an even more complex environment, and figure out how in the world to manage this exponentially more complicated dynamic. The rules change according to the situation, but the stakes always stay high. They get highest of all when you raise things up to a level at which it's not just businesses, buildings, and highways that you need to keep operating smoothly and safely, but entire nations.

CHAPTER THREE

How does a single bullet start a world war?

Confused by Social Structure

THERE IS NOTHING IMMEDIATELY impressive about the community of stump-tailed macaques living at the Yerkes Primate Research Center in Lawrenceville, Georgia—apart from the fact that they've formed a government. You don't notice a government at work just by looking at the macaque troop. The twenty-pound, thirty-inch animals spend their days the way most macaques do—grooming, playing, eating, mating, and of course, fighting. It's in those fights, however, that the monkey government makes itself known.

Macaques are pretty good at choosing their leaders, and they do it by no less democratic a process than voting. Not just any macaque can stand for election, or even wants to. You have to be big, you have to be aggressive, and under all circumstances you have to be male. Fill those criteria and you can start your campaign.

Males in contention for leadership positions must be willing to assert themselves—taking the food they want and the mates they want pretty much whenever they want them. They have to be willing to threaten and strike when they're challenged for those resources. And they have to master subtler skills too. They must know which individuals to court by grooming, which need occasional intimidating, and how often that intimidation needs to be applied. Do this well and the votes start to come in, generally in the form of deferential pant-grunts, bouncing, or bows from the lower-ranking monkeys. Facial expressions count as a vote too. A silent display of bared teeth from a subordinate individual—the closest you're going to get to a smile in the macaque world—is one more sign that a supporting ballot has been cast. Enough of these and you're the boss.

Once a macaque has been installed in office, there's more than just glamour to the job. Large troops of macaques like the Yerkes community—which is composed of eighty-four members—often require three or four governing males, something closer to a ruling junta than a presidency. When a brawl breaks out between lesser-ranking males, the nearest of the dominant animals will drop what he's doing, walk over to the combatants, and stand at his full height. The combatants, often as not, will break off the battle and separate. The dominant male will then groom the closest of the two submissives. Next, he'll turn to the other one and groom him too. Finally he'll retreat and the two former adversaries, knowing what is expected of them, will proceed to groom each other. The truce thus brokered, the fight will end.

There are benefits to having a well-policed macaque troop that go beyond merely ensuring a little peace and quiet. In orderly groups, members tend to mix more freely, which leads to fewer bands and cliques and reduces the likelihood of turf wars among those subbands. Good behavior by individuals tends to radiate out across the group, with macaques that have just shared a grooming session showing a greater tendency to turn and groom others; those individuals then turn and groom still more. This leads to a sort of peace ripple across the entire community. Peaceful communities are happier communities, breeding more, eating better, and living longer.

Not all macaque troops are so lucky. Too few dominant males or too few talented ones leads to all manner of societal breakdowns, as Yerkes researchers have proven in experiments in which they've temporarily removed the highest ranking animals from the community. Without the linchpin of a leader, fights can ripple from a center point as quickly as grooming sessions do. There is less play, less physical contact while sitting, and more frequent formation of hostile bands. This can eventually lead to the disintegration or dispersal of the entire troop. It's only with the right balance and the right leaders that a macaque community can achieve a state that, for them, amounts to peace and prosperity.

"Macaque society is extraordinarily complicated," says biologist Jessica Flack, who studies complex social systems at the Santa Fe Institute and has helped conduct some of the leader-removal studies at Yerkes. "They're conscious animals with overlapping and sometimes clashing interests, capable of three-way interactions. And they have a complicated status signaling network, one that communicates both what the relationship is between two individuals at the moment, and what it will be in the future. This leads to a power structure, which leads to societal structure, which leads to social organization."

You could do worse. And the fact is, human beings—the most socially capable of all species—often do. In the multiple millennia since we scratched our way out of the state of nature, we've been struggling to come up with fair, stable, sustainable ways to govern ourselves. Sometimes we succeed; just as often, we make a hash of it. Yes, we developed sophisticated nation-states, but we also figured out ways for them to oppress their citizens and make war on one another. Yes, we were smart enough to come up with free elections, but we also invented ways to rig them. We're capable of the peaceful ouster of an unpopular government, but we're also capable of the military coup and the beer hall putsch. We dreamed up a good idea like democracy—which manages to bump along in its own imperfect way—but we also invented autocracies, plutocracies, kleptocracies, dictatorships, military juntas, fascism, communism, Naziism, petro-states, shiekhdoms, serfdoms, and others. That a sloppy system like democracy remains the best on our list of organizing systems says something about our paucity of good choices. That nearly every one of these lesser systems still exists in the world today says something about the trouble we continue to have learning from our worst ideas. We're misled by the very structure government provides to think that we've gotten good at creating it, but in fact we're still rank amateurs.

In fairness, of course, governing ourselves is hard, a job that herds together such diverse variables as economics, geography, information flow, and more. But those things can be built on some very simple rules. And it's those simple rules that make all governments—good ones, bad ones, struggling ones—go.

THE INFRASTRUCTURE OF a nation-state is a sprawling thing. Even small countries with modest populations—the Jamaicas, Benins, and Belgiums of the world—require a dense weave of civil institutions, including schools, hospitals, revenue collectors, power and water systems, an army, a postal service, and on and on. So complex a network of people and resources requires an enormous amount of effort to build and maintain. And while populations rightly applaud themselves when they succeed in raising a nation out of a one-time wilderness, they frequently give themselves more credit than they've rightly earned. Often, the fact that that nation was able to get founded at all is owed mostly to the luck of the geographic or climatological draw.

The early United States grew rich as much because of the industry of its first settlers as because of the friendly, deep-water ports that made the new country so hospitable to merchant ships. Great Britain became a world power thanks in part to angry waters and rocky coasts that discouraged military invasions. Africa may have suffered for centuries under the twin scourges of the slave trade and regional wars, but these are transient, if terrible, crises compared to the more intractable problems of poor soil chemistry and frequent drought. The magnificent rice terraces of Bali are successful because they are designed around the rivers that flow down the mountains, feeding paddies and farms and shaping an entire culture all along the way. And where would the economy of Panama be if not for its flukishly narrow midline that so lent itself to being snipped by industrialists from the United States who envisioned a canal cutting the jungle? Indeed, Panama would not even exist as an independent nation if Teddy Roosevelt hadn't recognized the value of the little ribbon of land and decided to stir up an already bubbling insurgency there, helping canal-friendly local leaders break away from Colombia, which then controlled the region.

Few regimes better illustrate the small things on which even big states can turn than the curious case of President Suharto, the dictator of Indonesia from 1967 to 1998. By most measures, Suharto was a standard-issue despot. He came to power via the usual military putsch (aided—as more than one nasty regime was—by the CIA, which apparently provided names of possible communist opponents, many of whom were then rounded up and killed). He held power with the usual arrests, detentions, and arbitrary executions. He criminally enriched the usual circle of cronies and amassed the usual multibillion-dollar fortune. What made him special was something he could take no credit for himself, but something he had the good sense to exploit to its fullest.

The Indonesian economy has always depended heavily on its wood supply, particularly its broad-leafed evergreens, or dipterocarps. Milled into timber, processed into plywood, pulped into paper, they provide the revenue that drives every other sector of the country. The problem with a wood-based economy is that it's awfully easy to cut timber and awfully hard to grow it back. Harvest too much and you murder the very forest system that keeps you going. In general, if you take much more than ten trees from a single hectare of tropical forest—about 2.5 acres—you do so much damage to the timber web that the whole hectare suffers. At least that's the case in most places, but the Indonesian forest is a lot heartier. Here, you can harvest as many as twenty-six trees per hectare and the ecosystem shakes off the loss.

What makes the tree web so extraordinarily resilient is something that is itself quite ordinary. Entangled with the dipterocarp roots is a tenacious fungus that weaves a dense, living mat throughout the soil. The mat not only holds critical water and nutrients in the ground for the trees, it also crowds out other, harmful fungi species. What's more, it provides a safe place for the trees' seeds to drop every year, protected from both the elements and foraging predators.The trees, in turn, provide the fungus life-sustaining carbohydrates from their roots. The result is that the trees thrive, the fungus thrives, and the timbermen who know how to turn a profit from that arrangement thrive too.

President Suharto may not have known a thing about fungus, but he knew an awful lot about the kind of power so renewable a resource as Indonesia's trees could bring. The year after he seized power, he seized control of all of the country's timber-producing land too. He then carved the forests into regions and handed the timber concessions over to his generals. The generals got rich, the military stayed loyal, and Suharto ruled largely unchallenged for more than three decades. The entire structure of the Suharto dictatorship, at least in its early years, teetered on the unlikely behavior of one of the simplest organisms known. Kick out the fulcrum of the fungus, and the new regime might have crashed and vanished before it even got started.

"It was that ecological system that allowed the political system to rise up in the first place," says Lisa Curran, complexity theorist and professor of forestry and environmental studies at Yale University. "There were Indonesian military men, Chinese timber tycoons, international exporters, and local politicians all profiting from this complex natural web."

Not all countries are perched on as precise a point as Indonesia under Suharto. It's mostly single- or dominant-resource states that find themselves so precariously situated—oil-dependent nations like Saudi Arabia or Venezuela, diamond-dependent states in Africa, tourism-dependent states in the tropics. A shock in prices or a bad run of weather can destabilize the entire economy and, as a result, the goverment. Most countries strive for a better balance of stability and nimbleness, allowing them to adjust to changing circumstances and remain in operation. As in so many other cases, this depends on where their political system lands on the complexity arc.

Unformed societies or failed states—essentially anarchic societies—are very dynamic things, but also very simple things, moving in all directions at once without any meaningful structure or order. Totalitarian states, at the other extreme, are nothing but structure and order, but they're similarly simple to model, since the social and political lockdown they impose is fixed and unchanging. In one case, it's chaos; in the other, it's robustness. In neither case is anything terribly complex going on. It's at the top of the arc, where there's a functioning social system but one that's open to constant change, that you find the ferment.

It's not only multiparty states or Western-style democracies that occupy that sweet spot on the complexity spectrum, but they do occupy the highest perch. What really defines a complex state is less its particular bill of rights or election practices than the freedom with which its citizens can exchange information, even if that exchange is furtive or illegal. A Chinese or Iranian system that nominally forbids the free flow of ideas but can't eliminate satellite dishes or a relatively open Internet is complex despite itself. A Russian system that nominally permits such freedoms but at the same time is buying up broadcast stations and shutting down newspapers is sliding toward simplicity no matter how it presents itself to the world. The more any government, willingly or not, allows arguments to be made and minds to be changed, the more it achieves real complexity. And it's those kinds of governments that draw the most attention from the scientists who work at places like SFI.

Free-thinking individuals influence one another untold numbers of times a day. Advertisers study the phenomenon with marketing surveys; politicians study it with pre-election polls; behavioral scientists study it in the dynamics of the family and the workplace. One of the best and clearest ways to watch it all play out, however, occurs in the theater, during the familiar experience of a standing ovation. American theatergoers have lately lamented the cheapening of this previously rare audience display, an homage that was once reserved for only the most extraordinary of performances. These days, in the United States if not in Great Britain, it is becoming almost perfunctory. But how does any one standing ovation start? Audiences don't go into the theater planning to stand at the end of the evening, and the truth is, many would just as soon not. And yet they do.

University of Michigan economist and political scientist Scott Page thinks it all goes back to what he calls the social signalers. Somebody has to get a standing ovation going, and the better you can be seen, the likelier it is that you'll be that person. A standee in Row A will attract more imitaters than a standee in Row D, who will attract far more than a standee in Row Z. One reason, of course, is simply that when someone in front of you stands up, you have to stand yourself if you want to see the curtain calls. But by this reasoning, only the people whose views are obstructed should follow the example of the front row. Yet it's people all over the theater who respond to the cue. What's more, the phenomenon plays out more or less equally in theaters with poorly raked seats, where obstructed views can be a problem, and in steeper halls with so-called stadium seating, where sight lines are better.

"The people in the front rows are the audience's celebrities," says Page. "They're the visible ones so they're the ones whose example is mimicked. The people in the back rows we call the academics. They watch and follow. In an entire theater you may need no more than ten signalers to get a standing ovation going. The tail really does wag the dog."

But not all standing ovations are equal, even in the applause-happy U.S. After some performances, the mimicry is quick and uniform, with nearly everyone rising, and doing so almost at once. After others, the imitation is spottier, even with the most conspicuous signalers leading the applause. Academics and economists have burned no shortage of intellectual oil devising models to determine what distinguishes one audience from another. Is it the design of the theater? The lighting or acoustics? Ultimately, it wasn't the experts with their computers who figured out what was going on, but a group of students from the California Institute of Technology, puzzling over the problem and applying nothing more to it than common sense.

At some performances, the students realized, there are a lot of impulse buyers or late deciders who purchase solo tickets and arrive alone. At others, there are couples or families or theater groups who buy numerous seats or even whole rows. The closer the connections in any one group, the more likely one of its members will be to lead all of the others. A single member of a theater party who follows the lead of a front-row signaler can thus become a secondary signaler, cueing dozens of fellow members to rise as a group.

What happens in the theater happens in the outside world too. Societal trends—fads, philosophies, political movements—can spread with remarkable speed, picking up momentum the deeper they seep into the population. Every person who buys an iPod or switches to the Green Party is essentially an advertisement for the product or idea, making it likelier that the next converts will come along even faster. But the first few people who made the decision needed a little more persuading than the ones who followed, and this is where a signaler you know well becomes more persuasive than a signaler you don't.

"Ideas spread by imitation," says physicist Michelle Girvan, who studies complex systems at SFI. "If you have ten friends and two of them start wearing purple hats, you may decide to wear one too." If you have a hundred friends and the same two wear purple hats, you still may be likely to follow their lead since, after all, you've received the same two signals. But that general rule holds only if the amount of time you spend with those key signalers remains the same no matter how many other friends you have. If you start dividing your time equally among your hundred friends, you see less of the hat wearers and are thus less likely to imitate them. "The amount of time you spend with other individuals changes their ability to influence you and your ability to influence them," says Girvan. "If you're someone trying to get a purple-hat fad started, you don't necessarily want to sell your first hats to the most popular people in your market, who may have a great many casual relationships, but to the ones with fewer relationships but closer ones."

Purple hats may not trouble social researchers much. But trends like drug use, delinquency, and criminality do. Kids who attend high-crime schools are significantly likelier than other kids to become lawbreakers themselves; kids who observe a lot of drug abuse are similarly likely to experiment too. And when the signaler is not merely part of a child's social circle but part of the child's family, imitation becomes almost inevitable. Children with an older sibling who smokes are four times likelier to pick up the habit themselves. Girls with an older, pregnant teenage sister are four to six times likelier to follow her lead. Similar rules of proximity hold in models of epidemics and disease transmission too. "People studying the spread of sexually transmitted diseases often assume that every contact is equally weighted, which would mean that if everyone has the same number of partners they distribute the disease equally," says Girvan. "But that's not how things operate. Some contacts are more frequent than others. Some are riskier or more intimate than others. You have to tune the model before you make these kinds of determinations."

What works with hats, ovations, and viruses works

with the spread of political ideas too. Stanford University biologist Paul Ehrlich and Princeton University biologist Simon Levin have studied the metabolic rhythms of participatory democracies, looking for what it takes to persuade individuals to adopt a new idea or belief or abandon one they already hold. Using computer models and historical analyses, they have come to conclude that most people will resist changing their opinions until a particular number of the people around them have changed theirs as well. Just what this threshold is depends on just what the new idea is. If the issue is arcane like, say, income eligibility for the alternative minimum tax, the threshold is low. A few good friends with a well-expressed opinion will probably be all it takes to flip you to the other camp. If the issue is more basic and polarizing—polygamy, abortion, capital punishment—only the most vigorous persuasion and most frequent exposure to the idea will have any chance of moving you.

Once you do flip your view, the transformation is often dramatic—the philosophical equivalent of the so-called phase change at which water suddenly transforms itself to ice at 32 degrees or steam at 212 degrees. "Very small changes at the microscopic scale can lead to phase transitions at the macroscopic scale," Levin says. "Physicists work with this, but it also has a role in how ideas spread in society."

The very power of strongly held beliefs—their stickiness, as complexity researchers say—is what keeps the system pegged in place even during times of extreme political unrest. Indeed, as emotions rise and voices are raised, democracies are frequently less likely to descend into turmoil than to move into stalemate—very much like the left-right rift that opened up during the Vietnam era, or the partisan gridlock that has gripped the United States since the 1990s. Things become even more mired when the issues being debated turn on religion or ethnicity.

Levin observes that the more threatened a group that defines itself by race or spiritual belief feels, the more it will raise its threshold for tolerating new ideas, actually moving away from the compromise the circumstances call for and toward absolutism. This serves nobody's long-term interests, but it does help ensure the purity of the group's members and reduces the likelihood of assimilation—which at some points may feel more like annihilation. It's no coincidence that the American minorities that are the most marginalized are often the most traditional ones, resisting such signs of inclusion as intermarriage or mixed communities. It's no coincidence either that on the world stage, those most ancient of enemies, Arabs and Jews, often come just to the brink of resolution and then collapse into violent resistance: Cease-fires and treaties are followed by intifadas; peace overtures by leaders like Anwar Sadat and Yitzhak Rabin are followed by the murder of the peacemakers themselves. "This kind of backlash—whether domestically or internationally—drives groups back apart and helps preserve the idea of 'the other' that seems so essential to their identities," says Levin.

Secular political groups can usually be relied on to behave less counterproductively, but not always. You'd think that pro-gun advocates would yield on the most onerous weapons if that might be a way to induce concessions on less controversial ones, but instead they defend assault rifles and explosives as fiercely as they defend target-shooting guns. You'd think the ACLU would yield on some types of constraints on the First Amendment if that might be a way to achieve larger constitutional aims, but instead they defend neo-Nazis marching in the neighborhoods of Holocaust survivors in Skokie as vigorously as they do newspapers publishing editorials critical of the government.

Girvan is working on computer models that can replicate the way this polarization takes place, and the best model that fits so far is the one that also applies when people are haggling over a price. The salesperson who asks $30,000 for a new car doesn't really expect to get so much, and the buyer who makes a lowball offer of $20,000 doesn't really expect to pay so little. The hope is that they'll barter their way to a point in the middle, which is probably closer to what the car's really worth. Employers and employees negotiating over salaries typically use the same strategy and, indeed, often misuse it, offering outrageously low salaries or requesting outrageously high ones in the hope of skewing the final figure. The abuse of that practice in such highly paid fields as professional sports is what led to the increased reliance on independent arbiters, contractural referees who may not split the difference between ask and offer, but instead must choose one or the other. This helps keep both sides honest. The mediocre outfielder is less likely to request a wholly undeserved $15 million per year if he knows that will force the arbiter to choose the relatively modest $2 million the team is offering. Better to ask for a more realistic $4 million and hope the person making the decision sees things your way. The same holds true for a stingy team owner who makes an unrealistically low offer and then gets stuck with the player's much higher figure.

In the auto showroom or the arbiter's office, of course, both parties have an immediate, personal interest in the matter being resolved, and so even the most extreme demands are usually softened. In the political arena, things are rarely so personally urgent, and so the extreme bids only grow more so, with the center eventually being abandoned and the combatants simply shouting at one another across a widening divide. The complexities of moderation—a place where compromises are crafted and deals get struck—is cast aside for the simplicity of extremism, where nothing at all gets done.

"In the United States, this is partly a result of the two-party system," says Girvan, "since there's no third position to offer an alternative. But it can also be a function of any passionate, competitive system, all of which have a tendency to drift toward polarization. The extreme view eventually becomes the dominant one and things just get stuck there."

But things can get unstuck too, never more dramatically than in the case of human rights. Few things are more persistently sticky than the beliefs that lead to bigotry, and yet institutions like the South's Jim Crow laws do get overturned. One of the most powerful, if least likely, things that eventually topples them is mathematics.

Complexity scientists studying social reform often cite what is known as the hidden Markov model of change, a nod to nineteenth-century Russian mathematician Andrei Markov, who helped quantify the ways different systems—physical, chemical, numerical—evolve from one state to the next. Markov stressed that while some variables in any process or model are visible, others are always concealed, though the ones you can't see are inferrable from the ones you can. This, of course, is a somewhat self-evident statement of how the world works. A freshly set dining room table tells you that a meal is being planned; a table full of newly dirtied dishes tells you one has just been eaten. But the principle applies in other, more complex ways too.

Markov-based processes, for example, are critical when you're learning a language or when software engineers are developing speech-recognition programs. Becoming adept at reading and speaking means hearing or seeing just a bit of a word or phrase and being able to draw inferences about what's around it. That's what allows you to scan a page of text quickly and understand whole thoughts from just a glance, or figure out the garbled speech on a bad recording in which many sounds and syllables are inaudible. Distill these probability-based hunches down to algorithms and teach them to a computer, and you have a mechanical brain that can hear and understand spoken speech—even if it can't yet do it nearly as well as a human brain can. Markov modeling has similarly helped scientists working on such exceedingly complex projects as disease epidemics. And it can also explain social change.

University of Siena economist Sam Bowles, who heads the behavioral sciences program at SFI, believes the death of South African apartheid offers the best possible example of how hidden Markov processes drive politics. In all oppressive social systems, people on both sides of the bias divide seek short-term safety and survival, even if this means long-term strife. For both oppressed and oppressors, then, the most sensible choice—what economists call the best response—is to accept the existing social order. Some people in the oppressed group, however, will choose to deviate—becoming non–best responders—and, as expected, lose the payoffs the system provides. This may mean social ostracism, loss of employment, or even imprisonment. The oppressing group witnesses this and decides that its best response for now is to dig in and maintain the status quo. It's only when enough non–best responders in the weaker group decide to forgo the payoffs of the system and rebel against it that that system seizes up and quits working. Now the best response—the one that provides the payoffs—is to yield.

This is just what happened in 1990, when Nelson Mandela marched triumphantly from prison and later assumed the presidency of South Africa. The transformation seemed sudden, but the hidden conditions—the ones that allowed the events of 1990 to occur at all—played out much earlier. Civil uprisings and strikes in the 1970s led by early non–best responders convinced white leaders that the country could soon become ungovernable. This led to a softening of labor rules, allowing for better airing of worker grievances and even a half day off to commemorate the 1976

Soweto uprising, in which township protestors were killed by government police—a concession that would be akin to the government of China declaring a holiday on the anniversary of the Tiananmen massacre. This in turn was followed in 1985 and 1986 by secret trips by representatives of the South African business community to the headquarters of the anti-apartheid African National Congress in Lusaka, Zambia, to begin exploring the possibility of dismantling the crumbling white-rule social order.

The anxiety of the white majority by then was unmistakable. In a paper Bowles wrote with his collaborator Suresh Naidu, he quotes an official of the South African Bureau of Mines faced with the possibility of a strike by gold workers in 1987: "The political situation in the country was really dismal. The issue wasn't what level of increases we negotiated; the issue was do we survive or not? Who controls the mines, really?" A business executive involved in the negotiations also wrote: "The most important thing that both sides learned is that you must not underestimate the bargaining power of your opponent and his ability to hurt you."

Throughout all of this, no more than 11 percent of the nonagricultural black South African workers ever became non–best responders, and those who did often acted completely unobserved by the rest of their countrymen—the hidden part of the Markov dynamic. Still, the subtly changing atmospherics in the country were impossible not to feel, even if they were impossible to see, and this made the eventual undoing of apartheid a less sudden, more gradual thing than it seemed.

The same kind of Markovian incrementalism held true for the fall of the Berlin Wall. The telegenic smashing of the wall took place in November of 1989, but over the course of the previous three months, demonstrations against restrictions on travel and the East German regime in general had been growing from small gatherings of hundreds of people to massive rallies of hundreds of thousands. Erich Honecker, the leader of East Germany, was forced out of office on October 18, but the government that took his place was not yet ready to yield on the protesters' key demands. Finally, early in the evening of November 9, the new leaders folded, announcing that East Germans would henceforth be free to enter and leave the country as they pleased. At the press gathering announcing the change, an official was prodded as to precisely when the new, liberalized rules would go into effect. He answered: "Well, as far as I can see, straightaway. Immediately." The German people took him at his word, and within minutes the pickaxes were flying at the wall.

In both Germany and South Africa, such quiet gradualism seems self-evident even without Markovian theory. After all, no oppressive system ever gives up voluntarily without being worn down slowly by the resistance of the oppressed. What makes the dynamic equal parts mathematical and political is the way the pace of the resistance can be represented arithmetically and, when it is, the way it fits so easily into Markov models. That was the essence of the paper Bowles wrote with Naidu, and that is the essence of his complexity work.

"It's what the biologists call punctuated equilibrium,"

he says, referring to the way species may remain stable for long periods and then evolve comparatively suddenly in response to subtle environmental changes. "You saw it with the fall of communism and apartheid and you saw it in the U.S. when Rosa Parks challenged Southern segregation laws."

A MARKOV MODEL may allow complexity researchers to apply broad principles of arithmetic to human dynamics, but nothing quite mathematizes social issues like an election. The most complex elections are neither the chaotic ones in newly hatched democracies like Afghanistan in which dozens of parties descend in a swarm, nor the pantomime voting in autocratic countries in which there's only one viable choice on the ballot. Rather, it's in the two-, three-, and four-party elections that lie between these extremes of randomness and robustness. Some of the most paradoxical features of complex elections have been best explored by Stanford University economist Kenneth Arrow, winner of the 1972 Nobel Prize for his eponymous Arrow Impossibility Theorem. Arrow's counterintuitive thinking included the idea that given the vagaries of electoral mathematics, it's often difficult for voters themselves to know which candidate they prefer, particularly in an election with three or more choices.

The Arrow Impossibility Theorem has broad sweep, taking in equilibrium theory, growth theory, and the economics of information sharing—most of it in entirely new ways. His political work, however, has deeper intellectual roots, going all the way back to the early ideas of the Marquis de Condorcet, an eighteenth-century French philosopher, mathematician, and political scientist. Condorcet was intrigued by the way voters who are given a lot of electoral choices often end up with someone most of them don't want. Modern economic theorists like Arrow are fond of a simple graph in which they illustrate how difficult it can be for consumers to know with certainty which of three ice cream flavors they prefer, but the same graph can apply when the choice is among three candidates.

Take the American election of 1968, when Richard Nixon and Hubert Humphrey ran as the nominees of the two major parties and George Wallace mounted an unusually vigorous third-party campaign that actually succeeded in carrying five states in the Deep South. Imagine three states in such an election—say, Pennsylvania, Arizona, and Arkansas—and imagine that on election day, the first-, second-, and third-place choices of the voters broke down this way:

CANDIDATE	STATE *Arizona*	*Arkansas*	*Pennsylvania*
Nixon	1	2	3
Humphrey	2	3	1
Wallace	3	1	2

Each candidate has finished first in one state, second in another, and third in a third, so within these three states at least, the race is a draw. Now ask all of these voters who they would prefer in a two-way race between Nixon and Humphrey. Here, Nixon is the clear winner. In Arizona, his first-place finish beats Humphrey's second place, and in Arkansas, his second place beats Humphrey's third. It is only in Pennsylvania, where Humphrey's first place beats Nixon's third place, that Humphrey prevails. So Nixon wins two states to one.

In a two-way race between Humphrey and Wallace, Humphrey would win, with a second-place to third-place advantage in Arizona and a first-place to second-place edge in Pennsylvania (though that is not how the two candidates actually finished in Pennsylvania, nor in any other state that Humphrey won). In Arkansas, Wallace's first place beats Humphrey's third place, but Humphrey is still preferred in two out of the three states, and he thus wins overall.

Now, if Nixon beats Humphrey and Humphrey beats Wallace, Nixon beats Wallace, right? Not necessarily. In Arizona, Wallace's third place loses to Nixon's first place. But in Arkansas, Wallace beats Nixon first place to second place, and in Pennsylvania Wallace wins again, with a second-place finish to Nixon's third. So the voters overall like Nixon better than Humphrey and Humphrey better than Wallace; at the same time, however, they like Wallace better than Nixon as well. "You can talk about true preference in a three-way race only when you can say that if you eliminate a losing candidate, the winning candidate would be the same," says Arrow. "That's not what would have happened here."

This kind of oddity plays out all the time in electoral politics everywhere in the world, and while most American presidential elections don't feature three viable candidates, most presidential primaries do, not to mention many state and local elections. In all of these cases, the voters' aggregate choice may have nothing to do with the choices made individual by individual or election district by election district. U.S. presidential elections are muddied even further by the historical curiosity of the electoral college, which electorally seals every state off from the next, with the ballots of millions of local voters determining how only a relative handful of electoral votes are cast. Even without a complete breakdown of safeguards like that which occurred in the 2000 election, the electoral college can make things awfully mathematically dicey.

In 1960, John Kennedy comfortably beat Nixon in the electoral college, 303 to 219. But the popular vote was far closer, 34.22 million to 34.10 million, or 50.087 percent to 49.023 percent. A shift of a few popular votes would have given Nixon the popular vote, but Kennedy the White House. In 1888, that's exactly what occurred, with Benjamin Harrison defeating the incumbent Grover Cleveland 233 electoral votes to 168, but losing the popular vote 5.53 million to 5.44 million. The same thing happened to Samuel Tilden at the hands of eventual president Rutherford B. Hayes in 1876. Brooke Harrington, the professor of sociology and public policy at Brown University who also conducted the studies of investor clubs, calls the electoral college a "distorting mechanism," one that "doesn't so much tally people's preferences as average them. That does not lead to the most accurate result."

Candidates may pay a price for other, less quantifiable kinds of distortion, ones that occur not on election day, but in the campaign leading up to it. History is filled with examples of viable candidacies that became suddenly derailed after a single gaffe or poorly phrased remark. Michael Dukakis's bloodless response to a 1988 debate question about the hypothetical rape of his wife only confirmed the public's impression of him as a chilly technocrat. George H. W. Bush's seeming unfamiliarity with a supermarket scanner during the 1992 presidential campaign was, fairly or unfairly, entirely consistent with people's sense that he was a privileged patrician. A similar thing was true for Gerald Ford's 1976 insistence that Eastern Europe was not at that time under Soviet domination, a manifest untruth that appeared to confirm existing doubts about his intellectual wattage. And while Bill Clinton's inhaled-or-didn't-inhale fudging in 1992 did not upend his candidacy, it immediately entered the argot and has stayed with him ever since because it reinforced his reputation as a serial hair-splitter.

The power of all of these gaffes was very particular to the character of each of the men. Had it been a populist like Clinton who committed the supermarket blunder it likely would have been passed off as a moment of confusion or lack of focus. Had a man like John McCain, with a reputation as a hothead, responded the way Dukakis did, it might have been seen as statesmanlike restraint. Though such uncharacteristic displays would reveal something unexpected about the candidates, none of them would have the newsmaking potential of the incidents that did occur, even though those incidents merely nudged the public's impressions of the men in the direction they were moving anyway. That little nudge, however, packed a big wallop. "Such moments animate an existing storyline," says Democratic political strategist David Axelrod, a senior adviser to Senator Barack Obama. "Simple things that confirm what people are already thinking about a candidate have greater impact than bigger, more surprising things that would cause them to reconsider entirely what they believe."

The same holds true for political parties. Democrats have spent the last forty years fighting a reputation as indiscriminate taxers and profligate spenders, particularly on such inefficient programs as welfare. Republicans have spent even more time being seen as the party of the rich and indifferent. Clinton's tough—some said pitiless—welfare reform program did little to change the image of his party. George W. Bush's generous—if still less-than-promised—spending on global AIDS programs has similarly failed to repair the GOP's rep. "Those tracks get laid," says Axelrod, "and it becomes very hard for any party to get credit for trying to shift its direction."

IDEOLOGICAL DIFFERENCES THAT play out within democratic systems can be exceedingly fierce, but there are also natural limits to them. Short of civil war or violent insurrection, even the most polarized democratic societies do not typically take up arms against their fellow citizens.

When the same kinds of differences play out between nations, however, things are very different, since nations always have the option of going to war. When they do, complexity theory often ensures they wage it badly.

The military is a characteristically stodgy organization. For all the talk of cavalry charges and lightning strikes, most armies are surprisingly slow and lumbering. To be sure, they're very good at a handful of things, particularly the mass assault, the beach-storming strategy of an operation like D-Day that relies on the principle that huge casualties—90 percent and beyond—are acceptable in the first wave of an attack, provided that in the second, third, and fourth waves the level of carnage slowly drops. The sheer number of soldiers the attackers can throw at the beach ensures that some will get through, seize the cliffs, and silence the enemy guns.

This is a technique we unashamedly—and largely unknowingly—cribbed from nature. Oak trees in the jungles of Borneo, for example, enjoy the explosive reproductive success they do in part because unlike other trees, they don't drop their seeds annually. Rather, they hold their fire until a tiny six-week window in February and March every four years. When that stretch begins, the trees fairly carpet bomb the forest. Any single oak can dump as much as 435 pounds of seeds across the surrounding hectare, and since the fruiting footprints of all of the trees overlap, most of the ground gets multiple deposits from multiple trees. This crowds out other fruiting species and is also good for the forests' pigs and chickens—which gorge themselves on the quadrennial feast—as well as for the local farmers, upon whom the same pigs and chickens are more dependent during the long lean years in between, when they happily take to domestication.

What works in the jungles of Borneo works in a lot of other places too. Inundation is a common strategy for many animal species with exceedingly high infant mortality rates. Fish that lay hundreds of eggs could hardly support hundreds of fry, but they never have to, since predators, illness, and even the parents snap up the overwhelming share of them, leaving just the most robust clutch behind. Conception itself—particularly among larger, live-bearing mammals—is an even starker example of inundation, with hundreds of millions of sperm speeding out in competition for a solitary egg and all but a single one falling short.

The military undeniably refined this kind of broad-front charge into a science that sperm, fish, and tree seeds could never match. What trips our armies up is when such a storming is not an option and they must find ways to do battle with what is often a lighter, fleeter foe. Evolutionary scientists classify human beings and all other living things as either explorative or exploitative. Exploitative organisms are the more static of the two, creatures with fixed niches and well-established survival strategies. An exploitative organism is unlikely to try something evolutionarily new, preferring instead to stick with what it knows and exploit its environment for familiar resources. This is good for the individual or for the next few generations, since playing it evolutionarily safe prevents you from making adaptive mistakes. But it can be bad for the species, which may be slow to adjust to a rapid change in circumstances—as the dinosaurs discovered when an asteroid strike suddenly changed the Earth's climate 65 million years ago. Explorative organisms are a different story. They tend to seek new niches, mutate fast, explore new survival strategies when the opportunities present themselves. This can be costly in the short run, since any evolutionary innovation has a chance of failing, but over the long run it keeps the species flexible.

In general, it's the big, multicellular creatures like us who fall into the cautious exploitative category, and the microorganisms and very small multicellular creatures that are explorative. A single virus can mutate to a wholly different state during a single passage through a single body. A human or a horse needs hundreds of generations to make even a relatively small transformation. "To a virus," says biologist David Krakauer of SFI, "you're a law of nature since you change so slowly and it mutates so rapidly."

The same differences of changeability and scale apply not only to living things, but to social or commercial organizations—and that includes armies. Militias like Hezbollah or terrorist organizations like al Qaeda may have nowhere near the destructive muscle of a true national army. But nor do they have the mass and inertia that makes armies such confoundingly difficult things to get from place to place. An army that makes a tactical blunder and finds it's heading into an ambush often does not have the speed or agility to reverse course. A cornered insurgency, on the other hand, often just melts away, seeming to dematerialize and then reappear days or weeks later in a place that the fighting looks better. As a rule, the more troops and equipment you have to assemble, provision, pack, and move, the longer it's going to take even the best-trained military to wheel into position or dodge a wipeout.

Armies aren't merely slow to get to a battlefield, they're slow to change the tactics they use there when the circumstances demand. World War I was one of history's worst bloodbaths in part because of the simple lack of imagination with which it was fought. The surprise appearance of trenches and barbed-wire emplacements did not prompt military planners to shift strategies—accelerating tank development appreciably, improving aerial bombardment, merely changing angles of advance to sidestep fixed lines. Rather, they stuck with what they knew, continuing to send soldiers charging over the top to get mowed down by machine-gun fire from enemies lying in protected ditches. Vietnam saw a similarly stubborn strategy, with American soldiers trained in the tactics of the traditional battlefield never quite adjusting to a guerilla war in which invisible combatants lashed out and then melted back into the jungles and the villages. Troops arriving to fight a similarly confounding war in Baghdad faced similar challenges.

Comparatively ancient wars, even without a lot of heavy equipment and vast distances to travel, played out the same way. The battle of Agincourt in 1415, in which 5,900 Englishmen led by Henry V defeated 36,000 Frenchmen, not only provided Shakespeare with the most gripping speech in what might be his most gripping play, it also illustrated the lethal power of the little against the clumsy vulnerability of the big. The grossly outnumbered English—only nine hundred of whom were trained swordsmen—abandoned the hand-to-hand strategy they normally used in such engagements and instead turned to the at-a-distance power of five thousand archers. But archers must stand, plant, and aim, a pose that makes them easy targets for charging soldiers. Without a strong human defensive line to protect the bowmen, the Englishmen relied on the relatively new idea of palings, sharpened posts hammered into the ground and pointed toward the enemy. The French, slower to adapt, obligingly charged the palings, with predictably messy results. Second and third waves of Frenchmen then piled up against the impaled first. And those were only the attackers who were able to get to the British lines in the first place. Heavy rains in the defile at the edge of the woods of Agincourt had turned the ground to mud, trapping many of the armor-clad French before they went more than a few yards. Historical revisionists have cast doubt on exactly how much of a mismatch Agincourt was, with some suggesting the disparity in forces wasn't nearly as great as military lore suggests. Even skeptics, however, concede that the French were the larger, more complicated of the two armies, and that the stripped-down British used this against them in one of history's great acts of military jujitsu.

Military colleges have long studied upsets like Agincourt and conducted simulations and hypotheticals gaming all manner of such military mismatches. Contemporary computer models now make these analyses much more sophisticated, but perhaps the best model is the simplest one, a back-of-the-envelope exercise the military colleges call Colonel Blotto. At its most basic, Colonel Blotto starts with two armies with one hundred soldiers each and requires them to be divided among three fronts. There are a lot of ways to carve up your hundred, leading to a lot of different outcomes. One option is this:

ARMIES	FRONTS	*Front 1*	*Front 2*	*Front 3*
Army 1		40	40	20
Army 2		34	33	33

In this case, both divisions are sensible and both are defensible, but Army 1's is better because it produces two battlefield victories and Army 2's produces only one (assuming that superior forces always lead to a win, which the first step of Colonel Blotto does). Now imagine Army 1 has 150 soldiers and Army 2 has only 52. Army 2 will probably be wiped out—but maybe not:

ARMIES	FRONTS	*Front 1*	*Front 2*	*Front 3*
Army 1		50	50	50
Army 2		0	0	52

In this case, Army 2 acknowledges its weakness and chooses to concede the fights on Fronts 1 and 2 and place its entire bet on Front 3. Army 1, which played it safe and prudently, winds up losing a battle to an army it should have crushed entirely. The game gets far more complicated when you don't simply assume that size equals a clean win, and instead add variables such as the degree of casualties when two armies are closely matched—52 to 50, say, instead of a more uneven 33 to 20. And none of that takes into consideration how complex the game becomes when you add other armies and other fronts and such x-factors as supplies or terrain.

If there is any consolation for Blotto commanders who guess wrong or Agincourt generals who are humbled by a smaller force, it's that the military does have a learning curve. Countries and armies do change their strategies, though it's usually the losers in any one war who are most willing to try something new in the next. Thus the French, victors in World War I despite their reliance on the too-robust strategy of fixed fortifications and concrete-hardened borders, returned to the same static tactics in World War II and this time were steamrolled. Thus the U.S., losers in Vietnam, reversed course in the first Iraq war, winning with a fast strike by overwhelming arms and a very clear exit strategy. The fact that these lessons were apparently forgotten by the time of the second Iraq war in 2003 is more a function of the particular tacticians than the larger laws of history.

POLITICAL AND MILITARY scientists might bristle at complexity researchers presuming to impose a new discipline on their very old fields. And the complexity researchers themselves concede that they are new to this game. Still, it's tempting if not entirely scientific to see very good applications for what the burgeoning research is revealing. Our places of business, our social groups, even our families are, after all, their own small governments, which need to create structural order out of the disorder of individual needs and demands. The Arrow Impossibility rules that shape the results of multicandidate elections similarly influence how we construct our circles of friendships. Just whose company you prefer at any given moment is not an absolute thing, but is determined by such variables as who's available—who, in essence, is on the social ballot—and the venue in which your socializing will take place. One friend may finish first in the contest over who should accompany you to a ball game. That friend may finish last if the destination is a museum. And only one particular confidant may ever be in the running if the purpose of getting together is to discuss a deep or personal problem.

The hidden Markov modeling that drives social change and powers speech recognition software can also save lives, as when an epidemiologist reveals the hidden link between, say, a cholera epidemic and a pump handle. Similarly, it's a very good thing when Markovian dynamics help to overthrow a regime like apartheid, but adolescent rebellion in a family develops in much the same way as the non–best responders' movement. It does not happen at once, but is presaged by lots of small and often hidden clues. Parents may successfully apply the strategy that failed for the white regime in South Africa, concluding that judiciously applied authoritarianism is sometimes the only way to run a home. In South Africa, it was a good thing when the ruling power fell; in a family, it's essential that the authorities hang on.

Human beings have always been confoundingly quarrelsome creatures, given far more to conflict than to resolution. The fact that we ever settle any of our differences is a tribute to how deftly we can learn to manage them. The fact that we so often fail to do so is a testament to how much more we still have to learn.

CHAPTER FOUR

Why do the jobs that require the greatest skills often pay the least? Why do the companies with the least to sell often earn the most?

Confused by Payoffs

GRIPE ALL YOU WANT ABOUT YOUR job, you can at least be grateful you don't work as an operator in a customer call center. It would be hard enough sitting in a room with two hundred other people at two hundred other consoles, fielding complicated questions from irate customers all day long. It would be hard enough explaining to your callers why they had to spend twenty minutes answering computer prompts and idling on hold, only to have you come on the line and ask them the same questions about their account number, birth date, and mother's maiden name they've punched in three times already.

What would be less bearable—and more galling—would be that at any moment, a man at a computer screen hundreds or even thousands of miles away could punch a few buttons and know every move you were making. Paid by your employers to keep an eye on all their operators, he could invisibly check on you as you work, monitoring how long you'd kept your last three callers on hold, how quickly you'd been able to resolve their problems, what time you'd gotten to work that day, whether you'd been late coming back from your break. Widening his frame a bit, he could gather up similar information about your call center as a whole. Are you fielding your assigned number of calls each day? Are you achieving your goal of answering 80 percent of them within twenty-two seconds? If you're taking too long, your managers might be sacked. If you're actually going faster, you and a few other operators might be laid off, the thinking being that the company can save a little money on salaries and still meet the twenty-two-second target.

The idea that call center work is thankless work is hardly news. But the idea that it's also complex work, demanding a range of different talents, is. It's a job that calls for patience, marketing skills, time management skills, even empathic skills, not to mention a deep technical knowledge of a host of products and services and the database abilities to process orders on the fly and under pressure.

"You need someone who can do a whole lot of things very well and do them all at once," says Rose Batt, a professor at the Industrial and Labor Relations School at Cornell University, and something of a call center anthropologist. "There's a complicated mix of performance metrics at play, not to mention constant time pressure. Calls cycle in to each operator at a rate of one every three to five minutes"—or up to 160 in an eight-hour day. "No sooner are you done with your current one than the next one pops into your head."

Perhaps more surprising than how skilled a job telephone work is, is how comparatively unskilled the jobs far above it in heirarchy and prestige may be. Does the number cruncher who helped develop the restrictive warranty rules on a new TV set work harder than the person who must memorize such details for dozens or hundreds of products and then explain them to impatient people on the other end of the line? Does the loss specialist who decides that a health insurance plan does not pay for a diagnostic test that might prevent someone from getting sick have a tougher go of things than the person who must explain such rules to worried callers who bought the policy and now find themselves uncovered?

Up and down the job scale such oddities abound—the factory supervisor who needs greater managerial and interpersonal skills than the board member who helps run the company; the ambulance driver and paramedic who has far quicker instincts and far defter hands than the chief of cardiac surgery who hasn't actually stepped into an operating room in years; the taxi driver who knows his town's streets far better than the city transit chief ever could. Just which jobs we consider complicated and which ones we consider elementary, even crude, is a judgment too often made not on the true nature of the work but on the things that attend it—the pay, the title, whether it's performed in a factory or an office suite, in blue jeans or a gray suit. And while those are often reasonable yardsticks—a constitutional scholar may indeed move in a world of greater complexity than a factory worker—just as often, they're misdirections, flawed cues that lead us to draw flawed conclusions about occupations we don't truly understand. We continually ignore the true work people and companies do and are misled, again and again, by the rewards of that work and the nature of the place it is done.

IT WAS FREDERICK TAYLOR, the American engineer and efficiency analyst, who most memorably—and sometimes least artfully—tried to parse the division of human labor, in his landmark 1911 book, *The Principles of Scientific Management*. At the time, the steel industry was one of the American economy's principle engines, and Taylor wanted there to be no mistake about who did the real work in the critical sector. "One of the very first requirements for a man who is fit to handle pig iron as a regular occupation," Taylor wrote, "is that he should be so stupid and so phlegmatic that he more nearly resembles in his mental make-up the ox than any other type. The man who is mentally alert and intelligent is for this very reason entirely unsuited to what would, for him, be grinding monotony."

Taylor's thinking—such as it was—caught on throughout the industrialized world. A lot of it was sound stuff.

Motion analyses, for example, allowed him to determine how efficiently factories and other workplaces were configured and whether it was possible to save steps, time, and money by organizing them differently. A lot of it was micromanagerial nonsense—his belief in a "science of shovelling," with twenty-one pounds established as the ideal weight for a full shovel-load, leading in some cases to different-sized shovels being issued to workers depending upon the density of the material they were moving. The central flaw in his thinking, however, was his simplistic view about the complexity and dignity of certain kinds of work.

One of the best measures for judging the true complexity of a job is how easily it can be replaced by a machine. In the early days of the industrial revolution, the received wisdom was that technology would erode the workforce from the bottom up. The less art or intellect that was required in your occupation, the more likely you were to find yourself edged out by a machine. The factory, most people believed, would be the place this shedding of the human workforce would play out first. Line workers who spend their days tightening the same few bolts would be swept away by pneumatic machines that could do the job faster, more efficiently, and without complaint. Mid-tier supervisors would fare better, since no robot could manage the workers who would remain or make day-to-day production and scheduling decisions. The mere fact of the smaller workforce of manual laborers, however, would mean that at least some managers would be lost too. It would only be at the top of the organizational chart—where the big-brain, small-muscle executive jobs that could never be done by a machine were found—that jobs would be safe.

To a degree that happened. Robots did replace many bolt turners and metal cutters, but the attrition went only so far. Nothing could duplicate the practiced eye of an experienced worker who could notice a small flaw in a half-finished product heading down the line and intuit immediately whether this was a one-unit anomaly or a problem all along the route. No machine, similarly, could bring the multiple senses to the job that a human can, feeling the way a car door just doesn't click properly in its frame or hearing the faint rattle in a TV as it moves along the belt, indicating that something inside it was never tightened or seated right. Robots might replace the truly automatic, repetitive positions, but the jobs that required the high-speed feedback and communications skills humans bring to their work—which was a lot of the jobs—were safe.

Meanwhile, one level above the manual workers, the mid-tier managers did find their positions disappearing, as the self-regulating talents of the people on the line became more evident and the place to cut jobs and save money increasingly appeared to be among the supervisors who had been hired to oversee what was now a somewhat smaller, more independent workforce. At the top of the ladder, the bosses and executives, whose jobs often called for sophisticated understanding of markets and knowledgeable reactions to changing demands and trends, did more frequently hang onto their positions.

The computer revolution, which exploded in the latter

part of the twentieth century but actually got started decades earlier, blew an even bigger hole in the middle of the workforce by automating the handling not of materials or personnel but of information. This caused the midlevel job loss that started in the factory to migrate into the office, displacing people whose work involved such tasks as evaluating mortgage applications or insurance claims. While such a development may have blindsided a lot of hardworking employees who suddenly found themselves on the job line, it was in fact a very predictable result of basic complexity rules.

Like so many other things, the vast spectrum of jobs and professions follows a form of complexity arc, but in this case, the arc is turned upside down, forming something closer to a complexity U. At the left peak of the U are the bluest of the blue collar jobs, the ones often held in the least esteem and usually the most poorly paid. At the right peak are the whitest of the white collar jobs—very highly regarded and equally highly compensated. It's in the valley of the U that most people work, and it's also there—like it or not—that the jobs can be the simplest.

Frank Levy, a professor in the Department of Urban Studies and Planning at MIT, believes that nothing better illustrates how the complexity U works than the thankless job of the truck driver. If there's any work that enjoys less popular cachet than driving trucks it's hard to imagine one, and yet it's a profession that calls for an unusually complicated array of skills. Every moment a driver is on the road requires an elaborate processing of incoming cues—visual, tactile, auditory, cognitive—all of which must be followed by instantaneous, real-time reactions. It's not just that a computer or other machine would find it hard to manage the coordinated business of scanning, turning, braking, and accelerating the job calls for, it would find it utterly impossible to approach the intuitive reasoning a human driver brings to the job. Could a computer understand that a pedestrian in a crosswalk talking on a cell phone is likely to be more distracted than a person paying closer attention to traffic and thus requires a wider berth when you're turning through the intersection? Could a computer observe a baby in a stroller about to throw a bottle and forecast ahead to the next second

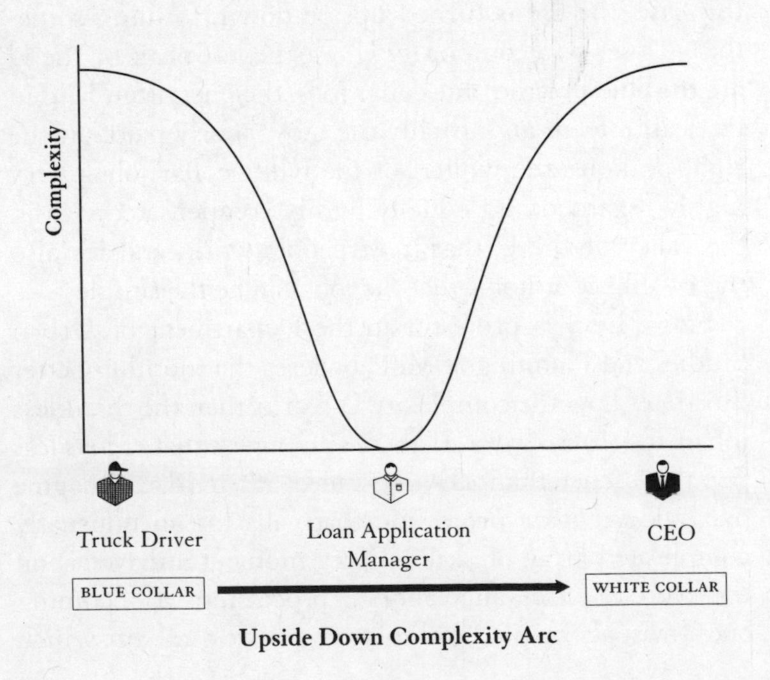

Upside Down Complexity Arc

when the bottle will hit the street and the mother will stoop to retrieve it, requiring the truck to brake and slow?

"Computers work well only when you can reduce things to rules and specify what should be done in all situations—the 'if A, then B' kind of processing," says Levy. "That's just not the kind of work a truck driver does."

Even lower-end jobs that already appear to have been swept away by computers are in fact doing better than they seem. No one spares much thought for airline ticketing clerks, workers who were never there in sufficient numbers to begin with and are now seemingly being pushed to extinction by automated ticketing kiosks. But the next time you're in an airport, take a look behind the counter. Even as fliers stream to the check-in machines, there seem to be just as many clerks as there ever were—and this makes sense. While a ticketing machine might be perfectly fine for the low-maintenance traveler with a single carry-on, it's no good at all to a wheelchair-bound passenger who needs assistance boarding a plane or the panicky parents of a college-age child who was supposed to be on an incoming flight but never showed up. Even the routine flier operating the machine always has a bailout option, simply by punching the "Agent" key if things get too complicated and joining the queue to talk to a live employee.

"If a customer hits a snag, human assistance is often the only way to solve the problem, particularly if the situation requires a little creativity," says Levy. "Humans, unlike machines, easily use something called case-based reasoning. We consider the current set of circumstances, compare them with situations that occurred in the past, and determine what we can take from those experiences. This is a kind of flexibility that's very hard to teach a machine."

Batt, who was impressed by similar improvisational talents among the workers at a health insurance call center she studied, added, "It wasn't just the eighty-five or so health plans that each operator had to study and learn, it was also the empathy and instinct they had to bring to the job. When a caller has just undergone an operation or had a child in a car accident, operators have to be able to listen well and counsel. This is thought of as 'emotional labor' and it's an intangible dimension, but a very important one."

The jobs at the other end of the complexity U rely even more heavily on those cognitive and instinctive skills, even as they lean less heavily on the physical ones. It's at this end that you find the lawyer wading through centuries of legal precedents to construct an argument for a current case; the pharmaceutical chemist parsing test results and other data and making an intuitive—even inspirational—leap that leads to a new cure or vaccine; the chef tasting a sauce, processing a stream of olfactory, tactile, and taste cues, and then plumbing a deep internal database for the one ingredient still needed that will transform all the rest; the psychologist listening to what a patient is saying, while simultaneously reading facial, vocal, and physical cues that reveal far more than the words themselves ever could.

It's only in the lowlands of the complexity U that things are a bit simpler. It's in this area, Levy explains, that the jobs most often involve the transmission of information up and down the line. In the United States, Europe, and other industrialized parts of the world, the growing ability of computers to do more and more of this kind of work has led to what the Japanese call *kudoka*—or a hollowing out—of the workforce. This is what claimed so many of Japan's fabled salarymen, the foot soldiers of the country's 1980s economic boom, and was one of the root causes of the long Japanese economic swoon, which ended only in recent years. The U.S. has been hit less hard than the one-time giant of the Asian economy, but the erosion can be detected in the American market all the same.

IT MAY NOT be much of a surprise that bosses and economic theorists don't fully appreciate the complexity in many jobs. Both groups, after all, are interested mostly in performance, never mind how it's achieved. More surprising is the fact that coworkers are often just as poorly informed about what the person one desk over or two spots down on the assembly line actually does all day. When they do give the matter any thought at all, they almost always conclude that the other person's job is far less complicated than it is.

Beth Bechky, a workplace ethnographer who teaches in the graduate school of management at the University of California, Davis, has extensively explored colleague-to-colleague job perception, wondering whether better informed coworkers might lead to better functioning companies. One of the first businesses she investigated was a semiconductor plant, and what she found there opened her eyes. When Bechky studies a workplace, she tries to become less an observer than a short-term employee, the better to plunge herself fully into the business. In a semiconductor plant that meant learning to work both in the engineering offices where the drafting of components is done and on the line where the technicians assemble the designs, taking instruction—even orders—in both places. This made it easier for both groups to let down their guard and reveal to her how their work gets done.

Engineers, Bechky observed, have what is essentially a two-part job: first dreaming up the elaborate designs that lie at the heart of the company's product, then turning those designs into the most detailed blueprints possible, so that the assemblers will have a precise map to follow as they work. When the engineers hand over the drawings, they do so with a simple, four-word dictum: "Build to the print." In other words: We've taken great pains to draft this just so, and as long as you do exactly what the schematics say, you won't bollix anything up.

The assemblers, in turn, accept the blueprints, shoo the engineers away—and often as not simply put the drawings aside. Never mind building to the print, many of them barely look at it. The source they turn to instead is one another. The more experienced assemblers figure out how the job should be done and tell the less experienced ones what to do. When a question arises, all of them simply consult one another.

"I sat in on meetings of the assemblers and they'd say, 'We're not going to use these drawings. Why do they keep giving them to us?' " Bechky recalls.

The answer isn't that the engineers are trying to patronize the assemblers by assuming they need such connect-the-dots instructions, any more than it is that the assemblers are trying to be obstinate by not reading them. Rather, it's simply that neither group ever took the time to find out the full scope of what the other one does all day. The designers thus never tap the builders' knowledge of the complicated materials they work with, and the builders similarly never learn all that the engineers could tell them about the finer details of their designs. When Bechky finished her research, she tried to make this fact clear to both groups, explaining the hidden complexity in all of the jobs at the plant. The entrenched anthropology of the place, however, turned out to be more powerful than the new lessons.

"I told them all what my findings had been," Bechky says. "A few months later I was making a presentation to the same group of people and repeated much of what I had said before. They were as surprised as if they'd never heard it."

If so technical a place as a semiconducter plant suffered from such complexity confusion, what about a more idiosyncratic workplace—like, say, a film set? Moviemaking is by definition a bohemian enterprise, one filled with untucked *artistes* unaccustomed to following a lot of corporate-style rules. As such it ought to occupy a spot far down one side of the complexity arc—the one that flirts with chaos on any given day. After her stint in the semiconductor plant, Bechky got herself hired as a production assistant—the least glamorous, most proletarian position possible in Hollywood—on both a low-budget horror film and a big-budget studio film. In this capacity she was able to move as freely throughout the organization as she had in the plant and take the full measure of the culture. Far from the free-for-all she'd imagined it might be, the film world turned out to be a buttoned-up, even inflexible, place.

With very few exceptions, a movie production is a gathering of vagabonds, a temporary organization brought together for what rarely amounts to more than a few months. Even stars and high-priced technical talent spend their careers jumping from job to job, sometimes working on three or more different projects in the course of a single year. The only way for such out-of-the-ether groups to come together and operate efficiently is for all of the people involved to know the precise limits and definitions of their work and be ready to perform it without instruction the day they arrive. That means the jobs must be fixed and predictable from movie to movie, with changes introduced only slowly and sparingly. "Much of what supports things in the film industry is a sense of historical structure that cuts across all of the jobs," says Bechky. "You just couldn't accomplish anything in a temporary organization without it."

This phenomenon, which students of the workplace refer to as "swift trust," is not exclusive to the movie industry; sports teams and the military depend on it too. In the film world, however, it's arguably more pronounced, not only because the organizations have such fleeting life spans, but because so much of the time is spent every day with a few people working—hanging lights, rehearsing a scene, dressing a set—and a lot of others standing around watching them. This leads to a better cross-job understanding and a better appreciation of other people's skills. What's more, film crews on location behave a lot like soldiers on maneuvers, bunking down in the same hotel or other accommodation every night and waking up in one another's company the next morning. This produces a total-immersion community that draws all of the members closer still. And should anyone start to drift, a culture of encouragement gently pulls them back. "Movie people remember to thank you for everything you do for them," Bechky laughs. "I'd be washing dishes and somebody would say to me, 'That's the most important job that's been done here today!' "

NO MATTER HOW simplified and streamlined a company gets, the system can still be blown to bits when the most confounding variable of all is introduced: the customer. Of all the organizational x-factors that most defy understanding, it's the consumers who actually attend the movie or buy the goods or services a company offers that are the hardest to parse. And of all the things that make that job so difficult, the first and biggest is figuring out what those customers want in the first place.

There is almost no limit to the product testing, consumer polling, and focus grouping manufacturers are willing to conduct in order to find the one product missing from an already overcrowded market that can become the next commercial blockbuster. The marketing landscape is littered with the remains of ideas that once seemed like sure hits until customers got a look at them and closed their wallets. Remember New Coke? Remember the McDonald's McLean DeLuxe? Remember the Lisa computer, a

$10,000 stinker from the folks at Apple, a company that today can do no wrong but for a while there seemed like it could do no right?

What corporations looking for the next iPod or Pringles or Cherry Garcia generally do is try a lot of things, often at once, hoping to find the one that sticks. If money is lost in R&D, it's made back many times over when one product becomes your jackpot. Few handicappers of the glutted beverage industry would have bet on Red Bull Energy Drink—a product way too sweet and way too caffeinated for many people's tastes—until it was actually released. These days it's available in more than 120 countries, with sales in the U.S. alone almost doubling annually in the first five years after its 1997 introduction, and climbing at a brisk, double-digit annual pace ever since. Chrysler's PT Cruiser, so ugly it's cute, could have been another bomb for the chronically ill carmaker. Instead, its cultlike following keeps producing new and return buyers, and while this alone might not be enough to ensure the company's overall solvency, it does keep people mindful of the Chrysler brand, bringing them into showrooms where they may be tempted by bigger, pricier models.

Deciding how many makes and models of any product to put on the market requires companies to navigate a decidedly treacherous complexity arc of their own, and where they land can spell the difference between big profits and bankruptcy. At one extreme, a shoe manufacturer may offer a single style of woman's sneaker in a single color, price, and size, and nothing more. At the other, the same company could make and market every known style of shoe from any known global culture throughout humanity's entire shoe-wearing history. The first case is the simplicity of unchanging robustness, the second is the simplicity of utter chaos. Neither is a sustainable business plan. True complexity and true profits are to be found at a critical point at the top of the arc, where you sell every type of shoe a meaningful portion of your customer base might reasonably want, and not a single shoe more.

Finding this spot is not easy, and not only because customers' preferences can be fickle. Adding a single product or feature to an existing line causes organization-wide complexity ripples. An automaker that offers a custom color like lemon yellow for a sports car has suddenly got to keep a lot of lemon yellow paint in stock. This ties up inventory space, revenue, and equipment. The same goes for glove-quality leather in the company's luxury car line, a chrome package in its SUV line, and any other detail on any other vehicle. What's more, it's not enough just to have all this raw material in stock, you've got to be able to mix and match it in an ever-multiplying number of ways. A television manufacturer that offers just five models differing only in screen size has a very manageable inventory. Now add three different speaker options—mono, stereo, or external—and you're up to fifteen possible combinations. Adding three different screen types—plasma, LCD, or traditional cathode tube—brings you to forty-five. And that's before taking into consideration all of the other electronic products and peripherals the company is likely to offer—from DVD players to stereos to cameras to cables and other peripherals—and all of the uncountable ways they too can be combined. In a 2005 article in the *Harvard Business Review,* Mark Gottfredson and Keith Aspinal, analysts with the global consulting firm Bain & Co., calculated that one large unnamed company that came to them for help had developed a line of products and options so huge that they could, in theory, be configured 10 billion different ways. Its nimbler and more profitable competitor had kept things to a still-complicated but at least fathomable three thousand.

"Our client's managers were unable to comprehend the operational implications of going from ten billion to three thousand configurations," Gottfredson and Aspinal wrote. "When we asked one of them what would change under such a scenario, he shook his head and replied, 'We only build a thousand units a day, so I can't think what would change.' "

But, of course, plenty things would change. Since you can never be sure which thousand configurations will be in greatest demand at any moment, you not only have to keep the material on hand to build the entire 10 billion, you've also got to teach your marketing people how to sell them, your repair people how to fix them, your distributors how to pack and ship them. It goes without saying that such excessive complexity means money wasted. "As the costs of managing that complexity multiply," warn Gottfredson and Aspinal, "margins shrink."

The answer, the two analysts suggest, is for all companies with multiple goods to sell to conduct a so-called Model T analysis. First, consider what you would manufacture if you were permitted to market only a single model of a single product. For Henry Ford in 1908, that was the Model T, a ten-gallon, twenty-horsepower car that came in only one color (black) and originally sold for $850, a daunting sticker price for a lot of folks at the time, but one that Ford was eventually able to slash to just $330 by 1927, the last year the car was produced—the equivalent of buying a new car for just $3,600 in 2006. For any contemporary company conducting a Model T analysis, the solitary product would be something different: For the shoe company it might indeed be a women's white sneaker; for Krispy Kreme it might be the plain glazed doughnut; for the folks at Campbell's it would surely be the iconic can of tomato soup.

Once you've picked your product, the next step is to imagine how you would make and sell it so that every possible penny could be earned from the process without sacrificing quality. Finally, you slowly begin adding back products, retooling and recalibrating the business plan as you climb the complexity arc, so that your operation remains lean and profitable while your line of merchandise grows. When you get to the point at which the cost of maintaining such a growing selection of products starts to erode the profits you're making from selling them, you've reached the top of the arc—what Gottfredson and Aspinal call the innovation fulcrum—and are in danger of tumbling down the other way.

The trick for all businesses is not just to locate this point, but to remain perched there. That is not an easy thing to do because in any dynamic industry—which is to say all industries—the innovation fulcrum is a constantly moving target. Take the case of Tiffany & Co., a perennially bullet-proof business that thrives during times of both boom and bust, if only because no matter the economic weather, there are always at least some very rich people ready to buy very expensive jewelry. But rich people will get you only so far in an era in which publicly traded companies—which Tiffany became in 1987—are expected to show constant growth.

If Tiffany were a snack maker under pressure to boost a sagging stock price, the answer might have been a new cookie. For Tiffany, the answer was silver—high-traffic items like key chains, picture frames, pens, and tie tacks that don't cost much but still attract shoppers with the promise of the company's signature turquoise box and white ribbon that signal a luxury gift. In the 1990s, Tiffany expanded its line of silver products dramatically and, just as it hoped, saw customer traffic increase and sales rise a whopping 67 percent between 1997 and 2002. That's good, right? Maybe not. The problem was, once the customers started showing up, they never quit.

At Tiffany, a place that made its name on white-glove treatment and deferential salespeople hand-selecting just the right item for you, a flood of customers buying mass market bling set all the wrong tone. Two thousand shoppers buying a $100 key chain each might mean a quick cash fix, but if they drive away a single shopper prepared to buy a $200,000 necklace, you've only broken even. Drive away two such high-end customers and you're in a $200,000 hole. Tiffany only made matters worse when it tried to deal with the sudden crush by issuing shoppers beepers as they entered the store, so that they could browse at their leisure and then be summoned when a salesperson was ready for them. Beepers are fine for people accustomed to patronizing a happy hour joint on a busy Friday night when the tables are crowded and you have to wait at the bar until you're buzzed. But that's not the preferred Tiffany customer—someone just as happy to stalk out of the store and go shop for jewelry at Harry Winston's, where they treat you like you've got real money to spend.

Tiffany had to act decisively and it did. First, it scrapped the hated beepers. Next, it did something few companies do casually: It decided to raise its prices, teasing what it charged for silver slowly upward a total of 20 to 30 percent, depending on the particular item. That scared off just enough lower-end customers to ease the crowding in the store, while maintaining a healthy traffic in the high-profit gift.

"Only after the company saw a dramatic plunge in satisfaction did it confront its fundamental managerial challenge," wrote Harvard economist Frances Frei in the *Harvard Business Review:* "Whether (and how) to serve two distinct segments of customers through a single retail channel."

The once-indestructible Gap similarly misjudged its market and has been trying to recover—with less success than Tiffany—ever since. The clothing company historically succeeded by limiting its choices to basics like button-downs, khakis, jeans, pullovers, and T-shirts, a stripped-down inventory that kept prices low and quality high. Yes, the Gap lost every single customer who was in the market for a suit or overcoat or any of the hundreds of other things it didn't sell, but it locked up the group interested in buying the basics. In 2002, the Gap rung up $16.3 billion in sales and was a fixture on city and suburban landscapes.

Jump ahead to 2007, however, and the company was stumbling badly, tripped up first by too much growth, with its expansion into the Old Navy and Banana Republic franchises, and second by simply giving its customers more than they wanted. Trying to compete with H&M and other bargain retailers that offer a wider variety of products to a younger audience, Gap began filling its shelves with a mishmash of styles and designs, taking a buckshot approach to its demographics instead of focusing first on the baby boom crowd that came of age with the brand. The company is fighting to regain its footing—indeed its very identity—and has considered spinning off the Old Navy and Banana Republic portion of its business and concentrating anew on the clothing basics. Whether such a return to minimalism will work is impossible to say, but what is certain is that the Gap paid a heavy price in customer loyalty by straying in the first place.

Simply manufacturing products efficiently and putting the proper ones on the market is not enough to protect companies from further complexity jolts courtesy of their customers. Frei has studied the manifold kinds of variability customers can introduce into an operation, particularly when that operation is a walk-in retail store. Customers show up at unhandy times; they ask for products you don't have in stock; they need a lot of assistance choosing things or learning how to operate them; they leave the place a mess, abandoning shopping carts throughout the parking lot or tossing unfolded clothes back on the shelves; they expect to be waited on the moment they enter the store—or, in the alternative, they expect to be left entirely alone until they're ready to summon a salesperson.

This kind of behavior can be brought into line in a lot of ways. Low- to mid-priced clothing outlets—including the beleaguered Gap—long ago learned to display pants in folded stacks in neat little cubbies, with the waist and length sticker prominently displayed. By now, most people know their W-34 L-35 as well as they know their SAT scores, and can go straight to the right shelves without assistance, simplifying the salesperson's day and reducing the number of employees needed in the store in the first place. Early bird specials, matinee prices, and holiday sales all train customers to come when you most want them, particularly if it's a time—like the dead 4:30 P.M. hour in a family restaurant—when the place is mostly empty. Customers, Frei points out, can even be taught to speak a company's particular language, an idea proven every day as Starbucks habitués master their ventis, fraps, and half-cafs, pleasing both themselves with their beverage erudition and the store managers, who know that lines move faster when customers aren't fumbling to describe a drink.

No one pretends that the vast, churning motor of the market economy will ever be anything but unfathomably complex. Like global climate, it can best be understood if you break it down to microenvironments—a single hardware store, a single coffee shop, a single fifty-person office.

Each has its own ecology, its own rich mix of interacting parts, its own sense of smaller order that contributes to far, far greater order. That, of course, raises a larger question: If the rules of business and workplace economics operate the same way at various scales, do other things as well? Can more of the institutions and structures around us be seen as a series of nested dolls within dolls, a succession of repeating structures writ small or writ large but all writ more or less the same way? They can and they often are, and within those echoing assemblages is found the next level of complexity, one of the most elegant of all.

CHAPTER FIVE

Why do people, mice, and worlds die when they do?

Confused by Scale

IT'S NEVER FUN WHEN A BUSINESS dies. Most do eventually, even if it takes generations, but that doesn't mean it's easy. A business becoming old and ill is just like a person or animal becoming old and ill. Its long-ago birth is far behind it. The growth spurts of its youth stopped years ago. It is fiscally frail, unable to withstand the jolts and shocks it used to take easily. Ultimately it is gobbled up by something bigger or simply closes its doors and quietly expires.

Nothing really dies when a business dies, of course, but

that doesn't stop us from treating it that way—with a sense of melancholy, even grieving. It's the same way with a well-loved car. The same way with an old house. The same way even with dying institutions or social customs: Politicos mourn the irrelevance of the party conventions; journalists mourn the shrinking of the network news; socialites mourn the death of the custom of dressing for theater.

Like it or not, all things die. Islands vanish; continents crack and recombine. Earth itself is a middle-aged world, born about 4.5 billion years ago and likely to stay around for another 5 billion or so, until the sun flares out and lights us up like a matchhead. Universities, libraries, cities all have life spans too, some lasting millennia, others just a few years. Tribes and communities spring into existence, endure for a few generations or perhaps hundreds, and then vanish. Most important though, to us at least, is the fact that we die too. The question we struggle with is not so much *why,* but why so fast?

The body is a lot of things, but at its basest it's an engine—a wet, self-regenerating machine that converts various forms of matter into various forms of energy, and keeps doing so for years and years. The day it quits converting energy is the day you live no more. But the tortoise is an engine too and it gets two hundred years. So is the redwood tree and it gets a thousand. The fruit fly, on the other hand, gets a month. All of those species—and us as well—are built from the same conceptual blueprint, and yet our countdown clocks run at radically different rates, meaning radically different life spans.

Is there some larger law that governs all of these things—trees, humans, institutions, worlds? Are the fruit fly and the city and the planet in some ways the same things—separated not so much by what they are but merely by how long they hang around? If so, the question then becomes: What determines that longevity? And the answer might be nothing more than size. Understand how small things scale up and large things scale down and you may just crack the basic riddles that allow them to function at all.

GEOFFREY WEST, THE president of the Santa Fe Institute, remembers the day he began thinking about heartbeats—specifically, how many he has left before he dies. He had just reached his fifties and begun contemplating the few decades that lay beyond when that mortal calculus occurred to him. As a mathematician and theoretical physicist he had never given matters of biology much thought before. Shortly after he arrived at SFI in 2004—just a few years after his heartbeat thoughts began—he decided that questions of life and death and size and scale might be where he'd want to focus his energies.

Well before West began his work, scientists already knew that heartbeats do not come cheap and that size plays a curious role in how they're spent. As early as the 1930s, Swiss-born chemist Max Kleiber began studying the relationship between an animal's mass and its metabolism, wondering particularly why smaller creatures burn energy fast and hot and larger ones operate at more of a metabolic simmer. Kleiber eventually came up with a formula that did not by itself explain the energy use but at least quantified it: For any creature, the amount of energy burned per unit of weight is proportional to that animal's mass raised to the three-quarters power. In other words, the smaller you are, the more calories it takes per pound or ounce or gram to keep you alive; the bigger you are, the more frugally you use what you consume. This explains why the smallest animals may regularly gobble many times their body weight in food in a single day, and larger animals, while they eat far more in absolute terms, consume only a fraction of the weight of their much greater bulk.

You can't, of course, eat as much and as fast as a small animal does without metabolizing it equally quickly, if only to make room for what you're planning to eat next. Flash-frying your food that way takes heat and energy, and heat and energy take heartbeats—lots of them, in rapid succession. It's not certain which came first—a heart rate that requires an abundance of fuel or an appetite that requires a castanet heart. Both may simply have emerged together, with their roots in the fact that small animals tend to be prey animals and thus must be quick. Whatever the reason, such frenzied little creatures also tend to live far shorter lives than drowsier ones, as if constantly running at so high a throttle simply wears them out early. It wasn't until the development of small and accurate sensing equipment that investigators could look more deeply into all this, measuring such things as body heat, blood pressure, calorie intake, and heart rate. When they did, the picture became clearer.

In general, all animals within a particular class—amphibians, birds, fish, mammals, or reptiles—get the same fixed number of heartbeats to spend in a lifetime. For mammals, the heartbeat budget is roughly a billion. A shrew, which weighs just twenty-one grams, or about four-one-hundredths of a pound, has a trip-hammer heartbeat of 850 beats per minute—and that's when it's relaxed. When the animal is frightened or fleeing, the speed can top out at 1,500. At that rate, the little shrew is dead within two years. Some species of whale, on the other hand, have heart rates that may be as slow as ten or fifteen beats per minute. When whales are diving, which they spend considerable time doing, they slow things down even further in order to conserve energy and limit the need to breathe. It's no coincidence that a whale living to be 100 years is not uncommon, nor that the oldest one ever studied was estimated to be 211 years old. On at least a few occasions, Greenland right whales have been discovered with broken harpoon heads in their hides that dated from a century earlier.

Up and down the mammalian class the rule applies, following Klieber's mass-to-the-three-quarters-power formula, with fleet creatures like gazelles getting just fifteen or twenty years and sluggish ones like rhinos getting almost fifty. The only exception to the rule is humans. At seventy beats per minute, pretty brisk by elephant or whale standards, we shouldn't live much past young adulthood. But our big brains—which themselves require lots of energy—have helped us game the system. Hospitals, medicines, and such extreme interventions as life-saving surgery—not to mention walling ourselves off from any natural predators and ensuring our food supply with grocery stores and supermarkets—have helped us push our hearts past two billion beats. Back in the wild, however, we'd average only half that.

West's first step in exploring this simple scaling law more deeply was not to conduct his calculations upward throughout the animal kingdom, but to factor down within individual creatures. If the Klieber ratio governs energy use from animal to animal, could it also apply to systems and subsystems within the body, perhaps even reaching all the way down to the cellular level? There was good reason to think it does.

If the body has a metabolic motor room, it's the cells; and if the cells in turn have their own little onboard engine, it's the mitochondria, where the Krebs cycle—the process by which carbohydrates are converted into energy—takes place. A sausage-shaped, self-contained organelle, mitochondria are thought to have once been free-floating organisms, before larger, evolving cells gobbled them up, seeking to enlist some of the mitochondria's prodigous energy-burning skills. The two structures have long since become interdependent, unable to survive without each other. But in terms of structure and function, they remain at least comparatively discrete.

Studying the metabolic rates of both tiny bodies, West found that just like two entirely different species, the relative rates of energy use indeed scale up according to the mass-to-the-three-quarters law too. What's more, the pattern of increasing efficiency continues as the cells collect into tissue strands, tissues into muscles and organs, and organs into a functioning organism. And not only does energy use within the subsystems follow the scaling law, so too does the very size of each subsystem. Capillaries grow into veins and arteries according to the same three-quarters-power scale. The same ratio applies to the way neural fibers become whole nerves and then nerve bundles. Taking all this together, from mitochondria, to cell, and all the way up the mammalian kingdom to the blue whale, the rule holds for twenty-seven orders of magnitude—or one times ten times ten times ten repeated twenty-seven times over.

"There's this exquisite interconnectivity," says West. "All of the structures have different forms and functions, but all of them adhere to the same scaling pattern."

WITH THE REACH of Klieber's law already extending wider than it was ever known to before, West wanted to determine if it could perhaps reach further still. It was here he took his first look at cities. It's never been a secret that living things and nonliving things operate on very similar principles. If there ever was any question on the point, it's been pretty well demolished by statistical physicist Eric Smith of SFI, with his remarkable—and exceedingly well-researched—theory that, never mind the old questions of whether human beings really descended from apes, the fact is we actually descended from rocks. Living things like us didn't emerge after the geological processes that allowed the planet to take shape in the first place, Smith concluded. Rather, we emerged *from* the geological processes—natural, even inevitable, descendants of some very simple science. "Metabolism is a complicated science," Smith says, "but it's also the one in which the rules are most like those of geochemistry."

Basic physical, chemical, and thermal systems are, at bottom, lazy processes, ones intolerant of stress or energetic tension. Blow up a balloon and the air will struggle to escape through the nozzle or a hole, seeking a relaxed state in which the pressure both inside and outside the rubber bladder are the same. Rain that collects on a mountaintop stores gravitational energy that it immediately seeks to dissipate by flowing down to the ocean, finding its way naturally into tiny folds in the topography of the hill. A few storms' worth of water erode the folds into a stream, which over time becomes a river. Nature abounds in such hunts for equilibrium—what Smith and others call relaxation pathways—and they likely shaped the very chemistry of the world.

In the Earth's earliest days, the planet's molten metals and other elements were stirred into a sort of undifferentiated goo. Radioactive heating close to the core and worldwide volcanoes higher up kept that primal matter bubbling and churning, and as it did, the various materials began to sort themselves out. Iron and other heavy metals fell toward the center of the Earth. Hydrogen, oxygen, and other light gasses separated from the heavy solids and rose to the surface, where they mingled with the early atmosphere. All of this stratified further over time, leaving us with today's layered world of deep, heavy metals; higher, lighter solids; and buoyant surface gasses.

"You have these big engines powered by nuclear reactions in the Earth's interior and they produce an energetically stressed environment," says Smith. "Those stresses had to be released and gravity found a way."

But even after the innards of the Earth organized themselves, the hunt for equilibrium kept things churning up above. Heat stored in the equatorial oceans after the long bake of summer upset the thermal balance between water and air. So nature invented great, convective storms that travel across the ocean, gathering heat as they go and then blasting it into the upper atmosphere—a process we came to call hurricanes. Evaporated water rising to the tops of clouds lost electrons along the way, which fell back and collected in the clouds' condensed lower regions. The negative charge created by the accumulated electrons led to a disequilibrium with the positive charge in the surface of the planet—creating, in effect, a battery, with the clouds and the ground serving as the plates. It would take a violent discharge of electrical energy to set the balance right, and so lightning was born. In a world built on electrical relaxation channels, gravitational relaxation channels, thermal relaxation channels, and more, it was inevitable that when biology emerged, it would follow the same rules too, and it did.

Perhaps the central pedestal of all biology—at least for life forms that rely on oxygen—is the Krebs cycle, named for Nobel Prize–winning physician Hans Krebs, who discovered the process in 1937. For any cell to survive, it must be able to convert carbohydrates into carbon dioxide and water and, in the process, generate the energy that sustains the organism. Krebs broke down the multistep chemistry by which this happens, and also discovered that it follows a telling pattern: The key molecular actor in the cycle is carbon dioxide, which passes through various stages, building more-complex molecules along the way and, at each step, releasing energy. The result of the cycle is remarkable—the survival of the cell and the larger animal it's part of. The means, however, is very simple: the hunt for relaxation pathways that will relieve energetic tension.

"The carbon dioxide just rolls down the energy hill, releasing energy and picking up carbon as it goes," says Smith. "It's just like the rain on the mountaintop." While the rain eventually leaves a riverbed in its wake, the carbon dioxide leaves the eleven basic organic molecules that serve as the precursors for all of the five hundred or so other types of molecules that make up every living thing on Earth.

The apparent similarities between rocks and organisms make it just as possible that other living and nonliving things overlap too, and cities are a natural place to look. By almost any measure, a city is humanity's most synthetic creation. We may have evolved to live our lives in a state of nature, but nature is also the place you get sick, cold, rained on, and eaten. The utterly controlled environment of a city solves all these problems. Cities are shorn of natural plant or tree life, except for the planned, symmetrical greenswards and parks we roll out like carpets. They're largely swept free of animal life too, with the predominant creatures able or willing to live with us—pigeons, mice, rats, bugs—hardly ranking among nature's elite. Our elaborate medical infrastructure even allows us to set up something of a screen against microorganisms, or at least push back at them when they close in. Yet cities also turn out to be among our most natural places, at least if strictly obeying the rules of complexity means anything.

There have long been plenty of reasons—well before Dr. Smith came along—to believe that urban centers are shaped by laws more powerful than mere municipal planning ordinances. There's the odd cellularity of a city glimpsed from an airplane window at night, a splatter of light resembling nothing so much as a brain cell, with the axons and dendrites of highways connecting it to the other cells—other cities—nearby. There's the vascular nature of highway systems—the biggest roads even given the name "arteries"—with individual vehicles bumping along like blood cells and the taillights flashing an aesthetically accurate red, a final detail that may not mean much scientifically but does appeal visually. New York City once kicked off a blood drive with a brochure that featured a map of Manhattan and the outer boroughs linked not by the familiar system of subway routes, but a circulatory system of blood vessels laid out in the same pattern as the tunnels. Not only did the image make instant graphic sense, it also made biological sense, with the more active and densely populated parts of the city more heavily vascularized than the outlying ones. If you think the same doesn't hold true in your body, think of the copious bleeding that accompanies even a small wound to your sensitive and well-muscularized face, compared to the relatively slow blood flow that comes from a deeper wound to the less-complicated flesh of the knee.

Maybe that's all just structural coincidence; maybe it's just the eye's tendency to impose familiar patterns in places they don't really exist. But there are other, more quantifiable rules that do seem to apply to cities and are harder to explain away. Take Zipf's Law.

In 1932, Harvard University linguist George Kingsley Zipf found himself wondering how frequently certain words and terms turn up in texts, and so, reasonably enough, he began counting them. Predictably, conjunctions, articles, and other workman words ("and," "an," "or," "a") appear most frequently. Highly particular words and terms ("zither," "glottis," "saline," "truculent") turn up least. What surprised Zipf was the pattern that governed that frequency. The most common word in any text, usually "the," will typically appear twice as frequently as the next most common word, usually "of." The third-place word, in turn, appears a third as much as the "the," the fourth-place word a quarter as much, and so on. Zipf himself couldn't explain why this was so, but in text after text, the pattern repeated itself. The phenomenon became even more of a puzzle when investigators noticed it appearing elsewhere as well, in comparative incomes, in sizes of corporations, and even, tellingly, in urban populations.

In 2006, New York City's population of 8.14 million people placed it first on the list of U.S. municipalities. Los Angeles's 3.84 million put it second, at 47 percent—nearly half—of New York's. Chicago's 2.84 million put it in third place, at 34 percent; Houston's 2.02 million was fourth at 24 percent; Philadelphia's 1.46 million was fifth at 18 percent. The list got muddled further down, since the sheer number of midsize urban centers caused four cities—Phoenix, San Antonio, San Diego, and Dallas—to cluster in a group of about 1.3 million residents each. But the dust clears at the number ten spot, where San Jose's 912,000 weighs in at 11 percent. Zipf's Law is clearly not perfect, but it applies often enough, certainly more than mere probability would suggest, that complexity researchers agree something is going on.

"We don't know why this is such a common power law," says SFI's Murray Gell-Mann. "But the fact that it's hard to explain doesn't mean it's not meaningful. We're already looking at simplified computer models to better understand why such a pattern would occur in nature."

If Zipf's Law applies to cities, could Klieber's too? West's early studies say that it does. Working with economists, physicists, and urban geographers, he has conducted at least rough counts of such indicators of urban scale as size of universities; number of public libraries; miles of roads, electrical cable, and plumbing; and even the number of theaters, gas stations, and hospitals. As the cities grow, he's discovered, they do tend to conform to the mass-to-the-three-quarters law. In the same way a larger animal uses less energy per unit of weight to sustain itself, so too does a city require less energy per building or block or neighborhood of growth. In the case of an organism, that energy is measured in calories; in a city it's in money or material or labor. A pattern that begins in the mitochondria turns out to repeat itself all the way up to Manhattan or Mumbai. Even human innovation seems to adhere to a similar—if not precisely the same—scaling pattern.

In 2004, investigators associated with SFI, Los Alamos National Laboratory, and the Harvard Business School published research in which they surveyed the number of patents awarded in the United States from 1980 through 2001, then sorted them not so much by cities, whose borders can be artificial, but by 331 so-called metropolitan statistical areas (MSAs)—somewhat more globular regions whose boundaries change and grow in a less regulated, more organic way.

In general, the investigators found that patent output among MSAs increases in what is known as a superlinear progression. Every time a population region grows in size by 100 percent, its patent output increases by 125 percent. If size and patents increased in a simple one-to-one lockstep, the reason would be simple: the greater the number of inhabitants, the greater the number of potential innovators, with no single inventor growing more productive than before. But gathering so many people together increases things more than arithmetically, concentrating talent and resources in such a way that it produces a so-called virtuous cycle, with a greater number of creative people leading to a greater number of shared ideas, which leads to a growth in institutions that foster creativity, leading to still more creativity, not to mention yet more creative people migrating in from other, less fertile communities. In this sense, innovators operate like individual cells growing together within a larger body. As they multiply, the larger organism of the metropolitan statistical area not only grows bigger, but more efficient, extracting more energy and output per individual component.

"Cities, despite their appearances, satisfy the rules of biology," says West. "The two things appear to be radically different, but it takes only a handful of comparatively simple laws to shape them both."

There are other, less calculable ways that organisms and urban centers overlap; the way they eat, for example. In general, fuel and nutrients move into and out of plants and animals in very predictable ways. Things that are useful in keeping the organism alive move from the macro level down to the micro, while things that are of little use or even toxic move back up and out. Food and water work their way from mouth to stomach to bloodstream to cell, where any utility is wrung out of them. What's left then moves back up into the bladder and intestines and is expelled as waste. Oxygen similarly works its way from mouth to lungs to alveoli to platelets and then back up to the lungs to be expelled as carbon dioxide.

That pattern, tellingly, is repeated in both the machines we build and in the communities we call home. Gasoline, coal, oil, and nuclear fuel move into the gullet of automobiles or the furnaces and reactors of power plants and eventually move back out as tailpipe emissions, smokestack exhaust, waste heat, and spent fuel rods. Similarly, bricks, mortar, steel, wood, glass, electricity, paper, water, food, and more stream into cities, entering by trainload, truckload, pipe, and cable, and filter down to homes, stoves, sockets, and people. The waste that's produced when these resources are used up or demolished is eventually flushed, hauled, and carted back into the world as trash, sewage, and factory emissions. Maybe we built our cities that way because we were unconsciously modeling nature; maybe we simply went with that template because the natural model is the only one that works. Whichever one it is, a single comparatively simple rule is nonetheless powerful and flexible enough to work in more than one environment.

Cities are similarly elegant—and similarly mimic the patterns of nature—in the way they adhere to what is known as the law of fractals. Nature is nothing if not redundant, with simple and familiar patterns playing themselves out again and again in both big and small ways. At the tiniest of scales, electrons circle atomic nuclei. At the largest, moonlets orbit moons, moons orbit planets, planets orbit suns, and great galactic arms wheel around galactic centers. Chemistry, geography, and botany work in the same repetitive fashion. Rivers branch, rebranch, and rebranch some more, forming the same expanding patterns at each level of division. Crystals, coral, bunches of broccoli—all follow a similar system of morphological redundancy, of smaller patterns within a succession of repeating larger patterns.

Humans have always known at least intuitively of this phenomenon. Scaled redundancy is evident in the stock market, where the minute-to-minute ups and downs of trading are repeated at least roughly in the day-to-day, week-to-week, year-to-year, and even decade-to-decade market charts. Ancient African hair-braiding and some forms of tile art are designed around such patterns within patterns, as have been countless forms of art that followed. The fractals name was not coined and the concept was not codified until 1975, when Yale University mathematician Benoit Mandelbrot first elucidated the idea of fractal geometry and transformed it from phenomenon to discipline. All the same, it's a phenomenon that's been with us since the beginning and it's one that, once again, cities obey. Electrical grids and municipal water systems begin at power plants and reservoirs, which diverge into feeder systems that divide and subdivide through a finer and finer web of wires and pipes until they reach individual rooms in individual homes.

The fractal nature of urban infrastructure may even help explain what many people consider the least pleasing feature of cities: the way they expand into every inch they're given. Urban planners know this phenomenon by the decidedly unlovely name "sprawl," but the fact is, in the world of complexity, such hungry growth may be unavoidable. Blood vessels and nerve endings divide and grow until they reach the very limits of an organism's extremities. If tissue is injured, new vessels and nerves quickly push back into the damaged area. A tree will stretch its roots to the limits of available soil; a houseplant will grow to the walls of the pot. Even nonliving fractal networks follow this space-filling pattern. Frost spreads to the edges of a windowpane as long as there's sufficient moisture and cold to support the formation of the crystals.

Fracture patterns in rocks and subsurface plates fan in all directions as long as the pressure that creates them continues to be applied. The human beings who build and occupy cities are simply following the same laws, sending the subterranean feeder roots of our pipes and wires and high-speed cables out as far as the available land allows, then raising houses and businesses and towns above them.

"Complex networks are space-filling," says West. "And nature is parsimonious when it comes to rules like that. It sticks to things that work, and reinvents only when it's absolutely necessary. Knowingly or unknowingly, we do the same."

None of this, however, means that any network is unlimited. The first obstacle is encountered at the micro end of the fractal spectrum. In all systems there are break points, irreducible levels below which it's not structurally possible to go. Shrews and whales are radically different in mass—on the order of 100 million, in fact—and that disparity continues deep into their anatomy. But down at the smallest possible level—the level of the cells—there is no size difference at all. Indeed, the cells of all animals great and tiny come more or less in just a single, democratic size. The same holds true for the smallest complex structures those cells make up: tiny capillaries, nerve strands, and muscle fibers, which are indistinguishable in size from animal to animal. In the botanical kingdom, the rule applies too, with the leaf cells and smallest water-transport vessels of the orchid differing very little from those of the redwood. And unavoidably, urban design follows the same dictum. No matter how large the city or town or building gets, every electrical system ultimately ends at the light switch, every plumbing system ends at the spigot, every room begins or ends with a door of a more or less standard size. All of these things are utilitarian devices defined less by aesthetics—which provide a lot of design options—than by the size and function of the human body that will use or operate them. "The terminal units," says West, "are optimized to the job they have to do."

So what does all this fractal redundancy matter? Aside from the fact that such wheels within wheels are somehow amusing, what do they actually teach us? Well, how to keep from destroying ourselves for starters.

One thing growing plants and animals appear almost to understand is when to quit growing, principally because they are calibrated to conserve their energy. The human body—the organic machine we know best—is a masterpiece of energy efficiency. For all the uncountable systems and subsystems your body must run to keep you alive, at any given moment it's burning only about one hundred watts of power. Head to toe, mitochondria to major organ, you're the consumptive equivalent of a single, moderately bright lightbulb. What's more, you couldn't exceed that much even if you wanted to. While the number of cells may increase linearly with mass—the overall cell count doubling as body size does—the energy those cells are able to extract from nutrients and oxygen doesn't. Capillaries and nerve fibers and other bodily subsystems can't expand indefinitely, growing further and further without limit, simply because the power to feed them doesn't keep up. When that happens, the tissue itself cooperates and quits reproducing itself. "The body's support for cells doesn't keep pace with the rate at which they multiply," says West, "so that multiplication simply stops."

But all that efficiency is maintained only within your body. Out in the world, you have different energy needs. This is called your social metabolic rate. Every light you switch on, every pot of water you boil, every newspaper you buy and discard, dramatically exceeds a modest one hundred watts. Indeed, figuring our consumption, purchases, and discards conservatively, each of us is averaging a whopping ten thousand watts. "When it comes to your social metabolic rate," West says, "you're acting less like a human being than a blue whale." To be sure, the analogy is not a perfect one, and a one-hundred-watt social rate is not a reasonable—or even survivable—figure. Even individuals in indigent cultures use at least a few hundred watts apiece. Still, as with so many other things, cultures in the West are helping themselves to far more than they need.

Such energy gluttony has a price, at least evolutionarily. Small animals that burn modest amounts of energy—albeit very quickly—may die fast, but they're also the ones that tend to reproduce the most, giving birth many times, usually in litters. Larger animals that require more calories to keep going are inclined to carry singletons or, at the most, twins. Human beings have never been litter-breeders, nor would we want to be, but across history, even our relatively modest fecundity rate has declined. The richer we've gotten, the more resources we've acquired to support children and yet the fewer offspring we've had per family. While the availability of birth control and an awareness of the dangers of overpopulation have surely been behind some of this, another, more primal factor is that we're simply directing our metabolic resources to consumption and growth instead of procreation and child-rearing.

"Even in a prosperous city," says West, "there's just a fixed amount of watts for each of us to use. We're burning too many per family, so we keep those families small. By this model, the best way to curb population would simply be to raise everybody's income level."

So what happens as our wealth increases even further? Are we headed for a childless future as mammoth, fifty-thousand-watt organisms? Not likely. While a civilization's desire to consume can grow largely unchecked, as with cells, the resources available to satisfy that drive almost always hit a wall. As creatures of impulse and appetite, we generally don't see the end coming—or if we do, we convince ourselves there are ways to dodge the problem. Exploding debt does not lead us to live more frugally, but to raise deficit ceilings. The soaring cost of gasoline doesn't persuade us to drive less but to drill more. Historically, we've behaved this way and, historically, one of two things has happened: The system has crashed, in bankruptcy, depression, war, disease, or something else; or an intervening innovation has reset the clock, giving the culture a chance at a relatively fresh start. The industrial revolution was one such game-changer; so was the digital revolution. Unfortunately, so too were World Wars I and II and the development of nuclear weapons.

In each case, the cultural landscape was gouged out and then replanted—with varying degrees of suffering or dislocation in the plowing, and various degrees of recovery after the resodding. In all of these cases too, West says, human civilization was not so much moving in a smooth, clean upward arc, but in a sort of scalloped pattern, with old cycles ending and new cycles beginning—sometimes happily, sometimes horrifically. "Inevitably," he says, "something has to collapse. It's the cycle of innovation, and that's good. But there's a price to pay for it."

Paying that price will always be an unavoidable part of growth and life. Remaining mindful of the fact that such a marker is out there, however, might help us delay the day that the debt comes due.

Happily, not every complexity-based choice we make in life need be so all-fired serious. Sometimes, the rules can be just as elaborate and the dynamics just as complicated, but the stakes can be so low as not to matter at all. It's in situations like these that we don't so much work the rules of complexity and simplicity, but play with them. And it's here that they can become fun.

CHAPTER SIX

Why do bad teams win so many games and good teams lose so many?

Confused by Objective

HERE'S WHAT HAPPENED ONE AFternoon in 2006 at M&T Bank Stadium in Baltimore, when the National Football League's hometown Ravens played the visiting Tennessee Titans: In the first play of the game, the Titans' quarterback handed the ball to running back Travis Henry, who ran forward mostly untouched for about fifteen yards. Finally, some Ravens caught him and wrestled him to the ground. The referee blew his whistle and the play ended. The Ravens eventually won the game.

Nothing remotely important happened in the twelve or so seconds it took to run that play, nor in the three hours it took to play the entire game. Nor, for that matter, did anything remotely important happen in any of the other professional football stadiums around the country where similar games were played.

The fact that the things that happened in the NFL that day were not important, of course, is not the same as saying they weren't extraordinary—or rich or elegant or artful or complex. Because the fact is, they were. The violent dance that is professional football is a high-speed ballet of many, many thousands of moving parts, coordinated far more precisely—even delicately—than the muddy scrum of an NFL game would suggest. What's more, the same ornate complexity can be found in baseball, basketball, hockey, billiards, polo, tennis, volleyball, cricket—indeed in virtually any physical contest in which two or more people go head to head for the symbolic prize of a victory.

Athletic competition, for all its seeming meaninglessness, has always been more than it appears. We're confused by the very straightforwardness of the objective to conclude that the playing of the game ought to be easier to explain than it is. But athletic competition runs far deeper than that. In most cases, it's a kind of kabuki affair, a consciously exaggerated pantomime of some very serious things: the ritualized chase of the soccer game mimicking the patterns of organized hunt; the advances and retreats of the football game mirroring the mortal tactics of the battlefield. It's not news that such similarities exist, nor that they're what give a sports contest its urgency, its ability to hold us fast in anticipation of an outcome upon which precisely nothing of real value turns. But observing those similarities is not the same as understanding them. If all human enterprises—warfare, economics, politics, the arts, the sciences—are built upon a few common rules and principles, the things that imitate those enterprises should rest on the same foundations. You can never truly get to the guts of sports until you get to the guts of those principles. And in them lie some very complicated—and some very simple—things.

TRYING TO PEEL back the complexity of a particular sport generally depends on which sport you're talking about. You can get a vigorous debate going anywhere in the world if you ask which of the hundreds of kinds of organized athletic competition is the most difficult to play. Almost without exception, regional favorites will top the local lists. Think lacrosse or rugby are tough to learn? Try buskashi, an Afghan favorite in which players on horseback do battle over the carcass of a calf. Consider kho-kho, a traditional Indian chase game in which nine players per team engage in elaborate pursuit of one another, following very precise rules and staying within a very precise playing area—just twenty-seven meters long by fifteen meters wide. The game lasts a maximum of thirty-seven minutes.

In the U.S., the question of which sport is the most complex is harder to ask, simply because there are so many games to choose from. There may be only four major sports leagues—baseball, football, basketball, and hockey—with a fifth one, soccer, slowly catching on. But everything from bowling to Ping-Pong to horseshoes to racing claims at least some partisan following in sports pages and on cable stations. Still, for American fans, it's the big five that win the most complexity votes, and the big two—baseball and football—that claim most of all. There are a lot of reasons for this bias, not all of them terribly scientific.

The more exposure we have to a game, the more we understand it; and the more we understand it, the more complex we like to think the rules are—the better to fancy ourselves connoisseurs of a very complicated thing. Every time a basketball or hockey game appears on the screen is one more chance to improve our scholarship of it. By contrast, the fewer opportunities we get to watch, say, a polo match, the more we can dismiss it as the effete distraction of some very unserious horsemen. We're especially vain about our grasp of domestically grown games, reckoning that while there will always be someone in the world who understands the sports of other cultures better than we do, nobody can truly fathom an American sport like an American fan. This is what gives baseball and football their edge over the other three.

Football's DNA may be rooted in rugby and baseball's in cricket, but their evolution in the United States has been so extensive and so particular that they have come to be regarded as a uniquely American species. Modern hockey is unmistakably a Canadian game and modern soccer European, and while basketball is arguably the most American sport of all, its back-and-forth style of play is so similar to that of hockey and soccer that it has an almost off-the-rack feel to it. Baseball and football just seem more like home-country sports, and we treat them with home-country favoritism.

There's more than just athletic chauvinism at work here, however. The fact is, there are structural qualities to baseball and football that may indeed make them harder puzzles to crack than most other games. For one thing, they take so damnably long to play, mostly because there's so much time in which no one's playing at all. The largely unpoliced rituals pitchers and batters go through between pitches and swings—adjusting their equipment, positioning their caps, stepping out of the batter's box, and backing off of the rubber—may be a venerable part of the sport's tradition, but hardly contribute to the action. Football's repeated time-outs—as players unpile themselves or ball position is determined or referees' rulings are disputed—have multiplied as TV has increasingly taken over the game and the action must stop even more frequently to accommodate commercial breaks.

Any game with such a deliberate, episodic structure simply allows for more strategizing and second-guessing—something that will always boost the complexity quotient. That's true not only of team sports, but of some one-on-one sports too. Jon Wertheim, a longtime writer for *Sports Illustrated,* and the author of a book on the culture and artistry of pool, has come to regard the people who play the game—or at least those who play it well—with true intellectual respect. There's a roguish, almost outlaw quality to the goings-on in pool halls, but the fact is, there's an enormous amount of deep thought and professionalism there too.

"There's this funny mix," Wertheim says. "The players tend not to have much formal education, but then you watch them play a game of nine-ball. They have to hit the balls in order for the shots to count, and by the time they're taking their opening shot, they're already thinking nine shots ahead. And it's not just a matter of setting up shots for themselves, but of putting the other guy in a position where he doesn't have a shot. The best analogy is to chess, which also calls on players to set things up many moves down the line."

Fast-paced sports don't lend themselves to this kind of planning and thought, and in fact, they suffer for it. There can be plenty of breaks between serves in a tennis game, but once a volley gets going, the ball may move in excess of 125 miles per hour—not a speed that allows for much reflection. "There's a lot of be-the-ball zen people talk about when it comes to tennis, a lot about the mental game," says Wertheim. "But if you ask Pete Sampras what was going through his head during a key point in a match, he'd probably tell you he was thinking about some song he heard on the radio that morning. A coach once said that the best thing about Björn Borg was that he never had two conflicting thoughts rolling around in his head at once. The truly great players can play the game without a lot of thought, or even much at all. As a result, they're inured to confusion."

But baseball and football are nothing but reflection and potential confusion, as players and coaches use the frequent breaks in play to practice the elaborate, three-dimensional strategizing that is as much a part of the game as the high-speed moments of action are. It's in that strategizing that the true complexity lives. So if baseball and football really are the most complicated of the five sports—something that is admittedly impossible ever to say with certainty—which is the greater of those two equals? Here, you'll get ferocious arguments, but the answer is actually pretty clear. One place to look is the rule book.

It's a basic principle of legislating and rule-writing that the more time you have to noodle with a body of laws, the more complex they will become. The federal income tax code began as a simple levy on money earned—not quite a flat tax, but something close. Today, it's a body of laws and enabling regulations more than 3,500 pages long. The most powerful amendment to the United States Constitution is the first, ratified in 1791, which ensures four primal freedoms—religion, speech, press, and assembly—and does so in an economical forty-five words. The twenty-fifth amendment, ratified in 1967, deals with the exceedingly narrow business of clarifying the line of succession when the president is ill or disabled, and requires four separate sections and 389 words to get that little job done. The constitution as a whole is a triumph of simplicity, able to fit on a single sheet of parchment when it was first drafted, and even today, strung with additional amendments, still containable in the stapled pages of a small booklet. The later-arriving constitution of New York State goes on for 45 small-print pages. California's, drafted later still, runs for 110.

Over more than a century, the elders of baseball and

football have been similarly prolix in their legislating. The official rule book of Major League Baseball is divided into ten chapters, and those are divided further into 123 subchapters. Chapter Two is devoted entirely to definitions, with an alphabetical list of terms from "adjudged" to "wind-up position." The entries run from the arcane ("quick-return pitch," an illegal throw made with deliberate intent to unbalance a batter) to the more self-evident ("touch," which defines contact with another player as touching any part of his body, clothing, or equipment).

Complex as this is, it's nothing compared to the byzantine business that is the NFL rule book, a document divided into eighteen chapters, two appendices, eighty-nine sections, and numerous supplemental notes. Unembarrasedly assuming constitutional airs, it even dubs its various subsections "Articles." Seventy-seven words are devoted to establishing the permissible size and color of the numerals on the jersery. A page and a half is given over to defining and describing the ball, which "shall have the form of a prolate spheroid," measure 11 to 11.25 inches long, and be inflated to 12.5 to 13.5 pounds per square inch.

While some of this specificity is no doubt excessive, the rules of football have good reasons to be more complex than those of baseball, not the least being that it's a vastly more violent game, requiring collisions to be vigorously policed, if only to prevent the players from killing one another. Take a rule on the simple chop block—a type of illegal, high-low hit in which two offensive players converge on a defensive player. There are ten different kinds of chop blocks specified, all of which can be murder on the man who absorbs them. Here's a description of number six: "On a running play, Player 1, an offensive lineman, chops a defensive player after the defensive player has been engaged by Player 2 (high or low), and the initial alignment of Player 2 is more than one position away from Player 1. This rule applies only when the block occurs at a time when the flow of the play is clearly away from Player 1." Imagine being a twenty-two-year-old recruit straight out of college and having to learn eighteen chapters of this turgid stuff. Then imagine being a coach having to make it all work on the field.

Hatching the strategy that opened the game between the Baltimore Ravens and the Tennessee Titans was the responsibility of Norm Chow, a man who knows a thing or two about how a football game operates. Once a player for the Saskatchewan Roughriders of the Canadian Football League, Chow quit in 1968 after a knee injury ended his brief career, then spent thirty-two years coaching college football at Brigham Young University, North Carolina State, and finally with the formidable Trojans of USC. In 2005, he made the jump to the pros as the Titans' offensive coordinator—a job that involves precisely what the title suggests: overseeing virtually everything that happens during the roughly 50 percent of the game that his team is in possession of the ball.

The complexities behind Travis Henry's fifteen-yard gain in Baltimore started the moment the eleven men per team lined up on either side of the ball to begin the play. A man facing you on the opposite side of the line, Chow explains, can generally assume one of three positions: He takes a stance off your right shoulder, off your left shoulder, or he can face you head-to-head. You can then line up in any one of three positions in return—switching to either of his sides or matching him face-to-face. This gives the two of you 3-squared total stances to present to one another, or 9. The man next to you can, in turn, assume three positions opposite his man, which triples the triple, giving the three of you 27 total starting options. The player across from him triples it again, for a total of 81. That quickly multiplies up to 243 when you add a fifth man, then again to 729 with a sixth. Since there are twenty-two total men on the field, the possible number of starting configurations before the play even begins is three to the twenty-second power—or 31.4 billion.

That's a lot of different ways for twenty-two guys to stand, and that's assuming each one of them has only three possible starting stances, which is not remotely as many as they actually do have, particularly because, unlike a baseball team, which must have the same nine positions filled on the field all the time, a football team may staff up its offense in a whole range of ways for each play, shuffling, say, the number of running backs and receivers it uses, just so long as the total number of men per side is never greater than or fewer than eleven. That flexibility exponentially multiplies the number of possible lineups.

"The offensive personnel shows changes from play to play, and every man you pull off the field and replace with someone else requires the other side to rethink its personnel too," says Chow. "Now start multiplying all those additional options."

What's more, every one of those players will in some way be involved in every play that's run in the game. This is not the case with baseball. In most baseball plays, the majority of players are permitted—indeed required—to opt out of the action. Even the sport's rarest and prettiest bit of fielding, the triple play, may involve only three fielders directly: the first, second, and third baseman. The other fielders must put themselves in position to run down a badly thrown ball, but they're only on standby. The double play may directly involve as few as two fielders, and similarly the putout at first. The fly ball nabbed in the outfield is the responsibility solely of the player in whose direction it soars. "In baseball," says Wertheim, "the majority of the guys aren't moving a muscle."

What's more, when baseball players do move, they often all do so in much the same way. With the exception of the pitcher and catcher—perhaps the most and least glamorous positions on the team, but in any event the most specialized—all of the players are required to master four basic skills: hitting, fielding, throwing, and running the bases. So common are the skills and so fungible the players, that team members routinely change positions throughout their careers—the third baseman who's lost a step moves to first; the aging infielder is reassigned to the outfield.

In football that doesn't happen. While a lineman may drop back to linebacker, or the odd running back throw the occasional pass, the sport is a sport of specialists, trained in one narrow job and sticking with it for a career.

"Part of the reason," says Wertheim, "is that you have such a disparity of mass. The three-hundred-pound linebacker bears little resemblance to the halfback who's half his size. They simply can't do the same jobs. That's not the case in baseball, where the physical profiles of the players are more similar."

That position-by-position specialization is reflected in the complexity of each team's playbook, which Chow and other offensive coaches fill with carefully crafted plays designed around the skills of their own teams. The Titans' offensive playbook is hundreds of pages long—the exact number and precise plays being something of a state secret. Throughout the season, the players must learn and practice that whole tactical catalog, though for any one opponent on any one day, they will prepare far fewer. In the course of a single game, each team has about eighty-five opportunities to snap the ball and run a play. A coach thus selects only about one hundred plays to draw from for that day, the choices based on the strengths of his team, the weaknesses of the other team, which players are hurt, which ones are healthy, and myriad other considerations. Once the gun fires at the end of the game, the players must shake off everything they've spent the week practicing and start rehearsing a whole new selection of plays for the following week's opponent.

"Once," Chow recalls, "when I was coaching in college, we came up with the idea of putting footage of all of the plays we'd run in games and practices on a video disk, which we'd give the players to study in the off-season. I was recording the narration, and midway through I stopped and said, 'Any player who has actually watched this far, call me and I'll give you $100.'" The entire team received the disk; only one player ultimately called. "These are kids," says Chow. "Sometimes we had to remind ourselves of that. They couldn't possibly be thinking about this stuff as much as we were."

THE ALMOST MONASTIC way coaches and managers lose themselves in their work is one thing that has given so many of them the reputation of rough-hewn wise men. It's not for nothing that Casey Stengel—player, manager, or coach with eighteen different baseball teams—was dubbed "The Old Perfesser." It's not for nothing that Bill Walsh, who coached the San Francisco 49ers to three Super Bowl wins in ten years, was known simply as "The Genius."

But more and more, sports' councils of elders are being filled not just by field generals, but by scientists and empiricists too. In the 1980s, tactical statisticians first began joining the game, unpacking their Apple IICs and first-generation laptops in press booths and crunching baseball's ubiquitous numbers in a truly serious way. Around the same time, physicists—moved in equal measure by the biomechanics and ballistics of sports as by what all this could tell them about how to design better and safer equipment—started weighing in, digitizing and analyzing the movements of players and the performance of their bats and rackets and mitts. In recent years, complexity researchers have been taking their own close look at things. Not long ago, the Santa

Fe Institute cohosted a seminar in which it brought together scientists and other observers of sport to take a more multidisciplinary look at the field, trying to tease out some of the larger rules that drive the games. One of the best-known in attendance was Dean Oliver, author of the book *Basketball on Paper,* and an analyst for the Denver Nuggets.

In the late 1990s, Oliver was toiling away at an environmental engineering firm, which was a position eminently suited to his education: a bachelor's degree in engineering from Caltech and a masters and PhD in environmental engineering from the University of North Carolina. But Oliver also loved basketball, and in his free hours he would fool about with game statistics, enjoying how the numbers opened things up to him in a way merely watching the games didn't. His academic training had taught him the power of equations to simplify such jobs as hunting for oil or protecting groundwater, and he believed they could do the same with the equally disorderly goings-on on a basketball court. "You learn all these amazing things in engineering school," Oliver says. "And so many of them rely on mathematical and statistical models that apply in other disciplines too."

Oliver was so confident of his analyses that he began offering them to the front office of his favorite team—the Seattle Supersonics—explaining that they would help the coaches get the most production out of each player's particular set of skills. The Sonics liked what they saw and for a while paid him with free tickets to games. Finally, in 2004, he published his book, and like the athletes he studies, he went pro, signing on as a consultant first with the Sonics and then jumping to the Nuggets.

If there's one bit of dictum above all others that Oliver tries to impress upon the coaches and general managers he counsels, it's this: Your hunches lie; trust the numbers. That's not necessarily an easy lesson for coaches to accept, trained as they are in the idea that while anyone can read and master a rule book, only the true scholars of the game have the gut sense that allows them to pull greatness from a team. Bum Phillips, the plainspoken former coach of the Houston Oilers and New Orleans Saints, once paid Don Shula, the former coach of the Miami Dolphins and Baltimore Colts, the highest possible compliment: "He can take his'n and beat your'n," Phillips said. "Then he can take your'n and beat his'n." In other words, never mind the roster, it's Shula's innate genius that determines the outcome of a game.

But genius, Oliver believes, is in many cases the residue of nothing more than arithmetic. That's an idea he demonstrated shortly after he came to the Sonics. In the 2004 preseason, the team's first and second leading scorers suffered injuries serious enough to keep them on the bench for a then-unknown number of games. Somebody had to step up and fill their lead role in the early part of the season, and for both fans and coaches, the obvious answer was Vladimir Radmanovic, an aggressive, high-scoring forward with a marksman's eye for the three-point shot. But Oliver wasn't so sure. It was true that Radmanovic led all of the remaining Sonics in scoring, but a look at the numbers just one level deeper revealed that he also led them in shots taken—shots that included a whole lot of misses. In a game that went on indefinitely, that wouldn't be a problem, since all that would count would be the points that showed up on the board, and Radmanovic would clearly post the most. But a basketball game runs for only forty-eight minutes of playing time, giving the average NBA team only eighty to one hundred possessions. Waste too many of them in junk shots and you won't have enough left for the good ones.

Oliver instead went a little farther along the roster to guard Antonio Daniel. Daniel scored a little less, but he also shot a lot less. Give both men an equal number of goes at the basket and Daniel would likely come away with more points. It was Daniel whom Oliver thus recommended for the top spot—a decision that was vindicated by the eventual numbers: At the end of the 2004–2005 season, Daniel had scored 843 points, Radmanovic 741. "Coaches make these judgments in their heads or by instinct," Oliver says, "but instinct can be deceptive. I plot these things not just along scoring continuums, but along skill curves. Where people lie along those curves is not always what it seems."

Of course, a truly successful team needs more than just a high-scoring player; to win consistently it also needs a leader who brings out the best in the rest of the roster. The sports world is full of surly, keep-to-themselves athletes who put up record-smashing stats but contribute nothing to the social and interpersonal machine that is also a team. Measuring the way the skills that emerge in the clubhouse complement those on display on the court was something else Oliver wanted to figure out, and he developed a mathematical model that allows him to do that. The formula he uses is a sort of meta-metric that combines a range of hard stats, such as shots taken, shots made, shots blocked, assists, rebounds, and more. Then he folds in vaguer measures, such as how the team plays as a whole when a player is on the bench as opposed to when he's in the game, as well as which teammates he works well with and which not. Then he stacks up all of these variables—giving some more weight than others—and arithmetically crunches them down. What he arrives at is a single figure that essentially gives each player a thumbs-up or thumbs-down for each game. When he compares any one player's rating to the team's performance in that game, something revealing emerges.

"You look at the really great players," Oliver says, "and their individual thumbs-up or thumbs-down records pretty closely correspond to the team's overall won-loss record. This is not just a function of whether they score a lot of points or not, but rather, what the larger effect is that they have on the team. That player, in a very real way, bears a great deal of responsibility for whether his team wins or loses. To me, that seems like a very good way to determine a leader."

There are other situations in which numbers succeed where instincts fail. David Romer, professor of political economy at the University of California, Berkeley, and another attendee at the SFI seminar, spends his days teaching the arcane topic of systems analysis and dynamic programming and how they apply in such areas as inventory management and consumers' purchasing decisions. But he has also spent a fair amount of time contemplating the mysteries of the football field goal.

The question of when and whether to try to kick a field goal can be a surprisingly difficult one for even the most experienced coaches. The last play of any failed drive is usually a punt, a deep kick to the opponent at the other end of the field that won't net you any points but at least forces the other guys to begin their own drive from as far away from your end zone as possible. When your drive dies very close to the opponent's end zone, however, things are different. There's no room to punt, and your only choice is either to take one final shot at a touchdown—which would net you six points with an opportunity to kick an extra point—or opt for kicking a much safer field goal, which will get you three.

Even if you're as close as the two-yard line, a touchdown attempt is no sure thing, succeeding only about three out of seven times from that distance. By contrast, a field goal from so close a range has a 98 percent success rate. Assuming you're not at the very end of a game and trailing your opponent by four points or more, it would thus seem to make sense to go for the sure-thing three instead of the gambled seven, abiding by the football dictum "Don't take points off the board." But hold on.

If you go for the touchdown and fail, the other team gets the ball, beginning their drive from the same spot, which is to say their own two-yard line, or ninety-eight very long yards from scoring themselves. If you play them tough and keep them from advancing very far, they'll have to punt the ball back to you from a spot still deep in their own territory—"in the shadow of the end zone," as the football coinage goes—meaning you're likely to begin the drive that follows from around midfield or closer, which is a very good place to start. This significantly increases your chances of scoring a touchdown on this possession, even if you came away empty-handed on the last.

Romer crunched the numbers from nearly every NFL game played in 1998, 1999, and 2000 and found that over the course of a full season, for every sure field goal you give up, you eventually get at least 1.4 points back in a subsequent touchdown. That's less even than a field goal, but it's more than the zero points it seemed to be. Such in-the-long-run, fractional scoring does you little good in the heat of a game if you need three full points here and now, but it does suggest that if you're patient and forgo the field goal when things aren't urgent, in many cases your touchdown will eventually come. "As a rule," Romer says, "teams should be much, much more aggressive on fourth down than they currently are."

Digging into the arithmetical roots of a game this way also reveals a surprising overlap between the rules governing sports and those governing economics. Ken Massey, an attendee at the SFI gathering and a professor of mathematics at Carson-Newman College in Jefferson City, Tennessee, is known locally for his academic skills, but he's known nationally—to sports enthusiasts at least—as the developer of the eponymous Massey Ratings, one of the best-regarded computer models available for ranking teams and players in more than a dozen college and professional sports. Massey's analyses are watched especially closely by fans and others who follow the college football bowls.

Unlike professional teams participating in their sports' postseason play, college teams hoping for a bowl bid do not qualify simply by accumulating the most wins in their division. Rather, they must be voted in by panels of former coaches, players, and others—a far less precise process that relies more heavily on the general impressions the judges have of the teams. Massey thinks the selection process ought to be less subjective—or at least more quantifiable—than that. One way to simplify things, he believes, is to treat each of a team's wins and losses as if it were a financial transaction.

Imagine a matchup in the regular season between USC, which wins an awful lot of games, and Temple University, which doesn't. If USC wins, the victors haven't really gotten something all that valuable, since there are a lot of games lost by Temple on the market. This high supply drives the value of any one victory over Temple down. If Temple beats USC, however, it's a different matter. A loss by USC is not an easy thing to come by. When something's scarce—oil, diamonds, an in-demand toy—the value and price go up. As a result of its loss, USC must transfer some of its value to Temple—far more than Temple would have transferred to USC had it lost—and Temple's value goes up accordingly. In the real world, USC hardly pays Temple a cash penalty for losing a game, but symbolic value does change hands. This can—or at least should—have an impact on USC's bowl standing, as the voters begin to view the team as something of a lagging commodity.

What's more, as with a stock, which can be pumped beyond its true worth if buyers start to see it as a moneymaker or can enter a death spiral if it is perceived to be a loser, winning and losing teams can enter a self-perpetuating cycle. Winners get pitted against other winners the following year, driving their value up further if they continue to play well. Losers play weaker teams, which diminishes the value of the victories they do manage to accumulate. Successful teams also attract better recruits, which improves their on-field performance further and drives up their value even more. All of this creates an economy of sorts, even if it's in a currency other than dollars.

Massey believes there are lessons here about globalization as well. Within a sports league, wins and losses are a zero-sum thing. There are twenty-eight teams in major-league baseball and each one plays 162 games between April and October. In a full season in which every game is played to completion (not something that typically happens, due to games rained out and not rescheduled) that would mean a total of 4,536 wins and losses, with that final tally always divided evenly—2,268 victories and 2,268 defeats—since every game will have a winner and loser. On the surface then, this means an overall league of perfect and permanent mediocrity, a state that's locked in by simple arithmetic. To determine which teams are actually better, you have to look deeper inside the data.

Major League Baseball's two leagues—the American and the National—are both subdivided into east, central, and west divisions. Members of, say, the American League Central and American League East play up to 35 of their 162 games against one another each year. Every game played between the two divisions will still result in one win balanced by one loss. But how these results are distributed will not be so equitable. Maybe the West will collect 75 percent of the interdivisional wins; maybe the East will; maybe they'll split closer to 50–50. However things come out, the rough justice of the playing field will give you a clear sense of which division is stronger, valuable information for both managers looking to improve their teams and fans looking for division bragging rights.

Making such quality comparisons between the American and National Leagues themselves used to be much harder. For most of baseball's history, teams on opposite sides of the league divide never played one another at all except for the mid-summer All-Star Game—a meaningless exhibition contest—and the World Series, in which both league champs meet for a maximum of seven games. In 1997, this changed, with the introduction of a system under which teams in each league play between fifteen and eighteen of their games against the opposing league every season. Purists wailed at yet another blow to the traditions of Olde Baseball, and their arguments had true sentimental merit. But they missed a larger point, at least if you're trying to get a real sense of which league is the better one.

"In the 2006 World Series," says Massey, "the St. Louis Cardinals beat the Detroit Tigers four games to one. If these were the only games between the two leagues, you'd conclude the National League was much stronger because they won 80 percent of the time. But that clearly wasn't the case." During that year's interleague play, the American League won 154 of the 252 games, or 61 percent. This was a cold slap of competitive reality for the National League, one that helped reveal the weaknesses of the various teams and target the areas that needed improvement. "A bunch of isolated conferences is like a bunch of isolated economies," Massey says. "If you don't allow them to mix, they stay primitive because you have no way of comparing them."

The more of that mixing you get, the more complex and even improbable the results can become, particularly when it comes to the way very bad teams can so often beat very good teams. It's axiomatic in almost all sports that on any given day any given team can beat any other given team. The Temple Universities of the world don't upset the USCs all the time, but surprise wins happen often enough to illustrate the way everything from weather to crowd noise to mere randomness can confound even the best handicappers. Even when bad teams don't beat good teams directly, they sometimes can by proxy.

The six-degrees-of-separation phenomenon—the idea that any one person on the planet is connected to any other person by no more than six other people—is an almost universally understood idea by now. Maybe your uncle fought in World War II. Maybe he had a platoon-mate who was born in Russia. Maybe the Russian platoon-mate is the nephew of a man who was an aide to Joseph Stalin. You're thus separated from Joseph Stalin by only four degrees. Massey has found that the same idea can apply to competition between two teams that are not playing one another at all.

The NFL's Carolina Panthers were hopeless in 2001,

winning only one game and losing fifteen. The Pittsburgh Steelers, on the other hand, were monsters, winning thirteen and losing just three. So can you forge a game chain by which the Panthers upset the Steelers? Easily. Carolina's sole win came early in the season, when they beat the Minnesota Vikings 24 to 13. Minnesota later trounced the Green Bay Packers 35 to 13. Green Bay went on to beat the Jacksonville Jaguars 28 to 21. And early in the season, Jacksonville shut down the Steelers 21 to 3. With only four degrees of separation then, an execrable Carolina team beats an excellent Pittsburgh team by a cumulative fifty-eight points.

Massey has reduced this hunt for such connections to an online application that allows him to search for similar upsets in any sport in almost any recent year. And while that's not much more than an absorbing distraction for fans logging on, for Massey it proves some very deep truths. "This is an illustration of the power of networks," he says. "You may have a relatively small number of friends, but they have friends and their friends have friends and all of those links are formed." By such links are economies, trends, political movements, and much more created.

The more meaningful upsets, of course, are the ones that happen directly, with the poor team beating the great team head-to-head. Such turnabouts are impossible for even the best-built teams to predict entirely, much less to prevent. Major-league baseball clubs from smaller cities have been gnashing their teeth for years about the overwhelming payroll advantage the New York Yankees enjoy, and with good reason. In 2006, the Yankees' single-season payroll was more than $194 million. The second-place Boston Red Sox spent $120 million. And the last-place Florida Marlins? Just $14 million, only 7 percent of what the Yanks' players vaccuumed up, and a little more than half of the $25 million the Yankees' Alex Rodriguez alone earned.

So the Yanks ought to win a lot of games, right? Well yes, to a point. But if ultimate victory in Major League Baseball is defined as winning the World Series, then the wealthy New Yorkers have a lot to learn from their struggling cousins in the provinces. In 2006, the Detroit Tigers had a payroll of $83 million, or less than 43 percent of the Yankees', putting them in fourteenth place on the money roster. Yet the Tigers went to the Series and the Yanks stayed home. In 2005, when the Yankees' payroll was a stunning $208 million, the American League and World Series champs were the Chicago White Sox, at $75 million. In 2003, when the Yankees were defeated by the Florida Marlins in the World Series, the New York payroll was $180 million; the Marlins spent just over a third of that, at $63 million.

Disappointed Yankee fans, as well as fans of other big-spending, oft-losing teams, have a lot of things to blame for their woes. One of the least-discussed is Colonel Blotto, the same back-of-the-envelope war game that military colleges use to illustrate the critical value not just of having a lot of troops but of deploying them well. Having a three-to-one advantage in soldiers avails you little if you overload one war theater with them and leave yourself dangerously exposed in another. A similar idea governs the dynamics on the playing field. A team that pours a lot of its payroll into a couple of overpowering left-handed pitchers is in trouble if it happens to come up against an opponent whose batters feast on southpaws. A team that excels at what managers call small-ball—eschewing the drama of the home run for singles and doubles and creative baserunning that slowly pile up the score—can be shut down completely if it comes up against a team with an unusually strong defensive infield. Throw all the money you want at your small-ball hitters or lefty pitchers, if the other guys figure them out, you're sunk. The answer is to distribute the talent you do have around the field in as flexible a way as possible, and spend the money you have to acquire that talent equally cleverly.

"Ranking teams in terms of payroll is like giving people IQ numbers and thinking that tells the whole story," says University of Michigan economist Scott Page, who also studies social signaling and crowd communication. "It's just too narrow and simplistic."

A similar thing is true in one-on-one sports like boxing. George Foreman's biggest advantage in the ring was his ferocious power, something he exhibited in the two convincing whippings he administered to Joe Frazier. Like Frazier, Muhammad Ali didn't have Foreman's power, but he had something Frazier didn't: reach. He could land punches Frazier couldn't, so he was able to defeat Foreman simply by covering up, wearing him down, and then coming in for the kill when the moment was right. "Colonel Blotto helps us makes sense of how this could be," Page wrote in a paper he coauthored with University of

Michigan mathematics graduate student Russell Golman. "We need to know where strengths and weaknesses lie and how they match up in a particular competition."

Coaches and managers are especially attuned to the importance of aligning their strengths with their opponents' weaknesses when they enter a playoff series, in which small differences are magnified, since everything may turn on a single game. Here, it's not just how you play or whom you play, but the sequence in which you play that matters. Say you're in a four-team baseball tournament in which each team has slightly different strengths in terms of three variables: hitting, pitching, and fielding. If 1 is the best ranking in each category and 4 is the worst, the competi-

TEAM	*Hitting*	*Pitching*	*Fielding*
Boston Red Sox	1	4	2
Washington Nationals	2	1	4
Chicago Cubs	4	2	1
New York Mets	3	3	3

tors might line up this way:

In his book *The Difference: How the Power of Diversity Creates Better Groups, Firms, Schools, and Societies,* Page cites a scenario very much like this and explains why it's a whole lot more complex than it seems. Say Washington and New York meet each other in the first round of the playoffs. Washington ought to win since it's dominant in two categories. Chicago would play Boston in their first round, and the Cubs would win for the same reason. This leaves Chicago and Washington standing, and Washington wins, based on its dominance in pitching and hitting. So is Washington the best team? Not necessarily.

Shake up the order of games and everything changes. Suppose Chicago plays New York in the first round. The Cubs win. That leaves Boston playing Washington, and this time Washington, the previous champion, doesn't even make it to the second round. Boston then plays Chicago and the Cubs win. If you shuffle it further and it's not Chicago or Washington that plays New York in the first round but Boston, the Red Sox emerge the champion. In this mini World Series, then, the only thing you have to do to win it all is to be lucky enough to draw the Mets as your first-round competitor.

No real baseball playoff is this simple, of course, and plenty of teams that are dominant in two or even all three critical categories wind up losing. But there's enough truth to Page's example that teams in all sports spend a lot of nervous days late in the season hoping not just to make it into the playoffs, but to line up against a team that will give them a better shot of making it to the second or third rounds. "So long as we know that the title of champion means 'winner of the tournament,' we're on solid ground," Page writes. "But we should not confuse 'champion' with 'best.' "

If something that seems as clear as the winner of a tournament is in fact as contingent as all that, the world of sports is a manifestly less precise, more fluid place than we ever imagined. What happens on the field has always lent itself to a wonderful statistical reductiveness, a kind of mathematical specificity that gives us the illusion that we fully understand the game we're watching. We don't, of course; the contests confound the numbers and the numbers confound our expectations often enough that we never fully crack all the secrets of any sport.

That, however, is where the true value of the games shows itself. The real worth of sports may have always been the way it pantomimes life. The vanity we bring to the sports we understand best—convincing ourselves that our favorite games are more complicated and nuanced than any other—is the same kind of preening we bring to other esoteric interests such as wine-tasting or ballet or fly-fishing. It may be true that all of these things have hidden dimensions that only true aficionados understand, but it's also true that some of those levels are illusory, functions more of affectation than fact. That's a useful slap of reality for the person about to spend $3,000 for a bottle of Côte du Rhone that you must convince yourself is worth the price or hundreds of dollars for a new rod that may not mean a thing for how much catch you bring home.

The ineffable value of a single leader of a basketball team and the way the qualities of that person go beyond the numbers on the scoreboard have lessons for every business or sales force ever assembled, as does the judicious risk-taking behind bypassing an easy field goal and going for the harder touchdown, gambling that the long-term payoff will be greater than the short-term gain. The greater complexity and subtler strategy to be found in a slower, untimed game like baseball similarly has larger value in a high-speed world in which standing back and taking a break allows a moment of clarity and distance that might not have been possible by simply charging ahead in a race with the clock.

Finally though, the true value of complexity in sport may come when you decide to quit looking for value at all. If, as Sigmund Freud may or may not have said, sometimes a cigar is just a cigar, so too a game is sometimes just a game. There's a reason that even in an era of big-money athletics, the operative verb we still associate with sports is play. That's one small way in which simplicity trumps complexity every time.

CHAPTER SEVEN

Why do we always worry about the wrong things?

Confused by Fear

IT WOULD BE A LOT EASIER TO enjoy your day if there weren't so many things trying to kill you before sundown. The problems start before you're even fully awake. There's the fall out of bed that kills six hundred Americans each year. There's the early morning coronary, which is 30 to 50 percent likelier than the kind that strikes later in the day. There's the fatal tumble down the stairs, the bite of sausage that gets lodged in your throat, the slip in the snow as you leave the house, the high-speed automotive pinball game that is your daily commute.

Other dangers stalk you all day long: Will a cabby's brakes fail when you're in the crosswalk? Will you have a violent reaction to spoiled food or the wrong medicine? And then there are the risks that are peculiar just to you: the family history of heart disease that claimed your father and grandfather in their fifties and could be shadowing you as well; the stubborn habit of taking chances on the highway that has twice landed you in traffic court and could just as easily land you in the morgue.

Trailed by danger the way we are, you'd think we'd get pretty good at distinguishing the risks that are likeliest to do us in from the ones that are statistical long shots. But you'd be wrong. We agonize over avian flu, which has killed only a small handful of people around the world so far, but have to be cajoled into getting vaccinated for common flu, which claims the lives of thirty-six thousand Americans each year. We wring our hands over the mad cow pathogen not likely to be in our hamburger and ignore the cholesterol that is—and that helps kill seven hundred thousand of us annually from heart disease. We avoid the water when we go to the beach for fear of a statistically unlikely shark attack, then fail to use the sunscreen that will protect us from a far likelier case of skin cancer.

Have we simply become overwhelmed by all the fear and uncertainty we face—remaining understandably jumpy over the 3,000 who died on September 11, for example, but giving little thought to the 220,000 who have perished on the nation's highways since? Looking askance at so benign a thing as spinach for fear of E. coli infection but filling our shopping carts with fat-sodden doughnuts and salt-encrusted nachos?

We pride ourselves on being the only species that understands the concept of risk, yet we have a confounding habit of worrying about mere possibilities while ignoring probabilities, building barricades against perceived dangers while leaving ourselves exposed to real ones. We micromanage the contaminants we allow to touch us, putting filters on faucets, installing air ionizers in our homes, lathering ourselves with antibacterial soap. "We used to measure contaminants down to parts-per-million," says Dan McGuinn, a former Capitol Hill staffer and now a private risk consultant. "Now it's parts-per-billion or -trillion."

At the same time, 20 percent of all adults still smoke; nearly 20 percent of drivers and more than 30 percent of backseat passengers don't use seat belts; two-thirds of us are overweight or obese. We dash across streets against the light, build our homes in hurricane prone areas, and when they're demolished by a storm, rebuild in the same spot. We allow ourselves to be talked out of sensible fears by manufacturers with products to sell—unhealthy foods, unsafe cars, cigarettes—and talked into irrational ones by politicians warning that electing their opponents will mean a nation debauched or endangered or worse. Rational calculation of real-world risks is a multidimensional math problem that sometimes seems entirely beyond us. And while it may be true that it's something we'll never do exceptionally well, it's almost certainly something we can learn to do better.

If risk is hard for human beings to parse, it's partly because we're moving through the modern world with what is, by many measures, a very primitive brain. We may think we've grown accustomed to living in a predator-free environment in which most of the dangers of the wild have been pushed away and sealed out, but as far as our slowly evolving central nervous system goes, this is a very recent development.

Joseph LeDoux, professor of neuroscience at New York University and author of the book *The Emotional Brain,* has studied fear pathways in laboratory animals and believes that much of what he's learned applies to human beings too. The jumpiest part of the brain—in both lab mouse and human—is the amygdala, a primitive, almond-shaped clump of tissue situated not far above the brain stem. When you spot potential danger—a stick in the grass that may be a snake, a shadow around a corner that could be a mugger—it's the amygdala that gets the message first, triggering the fight-or-flight reaction that causes you to jump. A fraction of a second later, higher regions of the brain get the signal and sort out if the danger is real or not. That fraction of a second, however, gives the fear response greater emotional impact than the rational response, and that advantage doesn't disappear. The brain is wired in such a way that nerve signals travel more easily from the amygdala to the upper structures than from the upper structures back down. Setting off your internal alarm is thus quite easy; shutting it down takes some doing.

"There are two systems for analyzing risk: an automatic system and a more thoughtful one," says Paul Slovic, professor of psychology at the University of Oregon. "Our perception of risk lives largely in our feelings, so most of the time we're operating on system number one."

This natural timorousness pays evolutionary dividends, keeping us mindful of dangers and thus likelier to stay alive and pass on our genes. Such timidity, however, is at constant war with an equally adaptive adventurousness—a willingness to take risks and even suffer injury in pursuit of prey or a mate. Our ancestors hunted mastodons and stampeded buffalo, risking getting trampled for the possible payoff of meat and pelt. Males advertised their reproductive fitness by fighting other males, willingly engaging in a contest that could mean death for one and offspring for the other.

Slovic and others say that all of this has left us with an admirably well-tuned ability to anticipate and weigh very clear, very short-term dangers. But across the entire spectrum of risk, things are often far more complex than that, as a little experience teaches us. For one thing, particular perils don't always spring at us from particular spots—a kind of unpredictability that makes avoidance learning hard. "If a tiger comes from a certain place one time, well then that's where tigers come from," says Nassim Taleb, a professor of uncertainty science—or, as he prefers to call himself, a philosopher of probability—at the University of Massachusetts. "You build a wall in a specific spot because the Germans invaded here once before, so that makes you feel safer."

The problem is, tigers won't always behave as expected; an attacking army may sometimes dogleg around your wall. The blissful inability to contemplate such novel possibilities keeps lower animals from worrying, but can also get them eaten or shot. Our more highly evolved talent for anticipating new scenarios allows us to take precautions against them, but also keeps us wringing our hands about all of the ones we may not have considered.

What's more, in a media-saturated world, you don't even have to dream up new dangers on your own, since newspapers, television, and changes in the government's terrorism alert from orange to red do it for you. It's in these situations, when the risks and the consequences of our response unfold more slowly, that our analytic system kicks in. This gives us plenty of opportunity to overthink—or underthink—the problem, and this is where we really start to bollix things up.

Which risks get excessive attention and which get overlooked depends on a hierarchy of factors. Perhaps the most important is dread. For most creatures, all death is created pretty much equal. Whether you're attacked by a lion or drowned in a river, your time on the savanna is over. That's not the way humans see things. The more pain or suffering something causes, the more we tend to fear it; the cleaner or at least quicker the death, the less it troubles us. "We dread anything that poses an increased risk for cancer more than the things that injure us in a traditional way, like an auto crash," says Slovic. "That's the dread factor." This in turn leads us to another, related distortion—overestimating the odds of the things we dread the most actually occurring, and underestimating the odds of the things we dread less. "It's called probability neglect," says

Cass Sunstein, a University of Chicago professor of law who specializes in risk regulation.

The same is true for, say, AIDS, which takes you slowly, compared to a heart attack, which can kill you in seconds, despite the fact that heart attacks claim fifty times more Americans than AIDS each year. We also dread catastrophic risks, the ones that cause the deaths of a lot of people in a single stroke as opposed to killing in a more chronic, distributed way. "Terrorism lends itself to excessive reactions because it's vivid and there's an available incident," says Sunstein. "Compare that to climate change, which is gradual and abstract."

Such sudden, aberrant events as terrorist attacks are what Taleb and others call the black swans of probability theory, a term that has a very long history. Until 1697, nobody in the Western world knew that swans could be any color but white; then Dutch explorers landed in Australia and discovered the black variety. The new birds seemed somehow exotic, even disturbing, but only because we had become acquainted with their light-feathered cousins first. "All of history is dominated by black swans, or exceptionally unusual events," says Taleb. "They all have three properties: They're very rare; they carry a huge impact; and after they happen we somehow concoct an explanation for them so that in hindsight we don't think they were so unpredictable."

Black swans that simply confound our expectations can be unsettling enough. Black swans that can kill us send us completely around the bend. There's no minimizing the traumatic impact of September 11 on the American mind, nor should there be. But knock down two buildings in

Baghdad or Beirut—even two such mammoth buildings—and the emotional blow would be at least a little less. There are simply so many other buildings that have been blown up or bombed into rubble in those tortured capitals that the public has become inured to the peril. That kind of habituation is true of all cultures; witness the aplomb with which Londoners came to handle the Blitz and later the V-2 rockets. Introduce a new kind of danger to any population, however—anthrax, a dirty bomb—and even the most imperturbable risk assessors get spooked. Not long after former KGB spy Alexander Litvinenko was murdered in London in 2006 with a minute amount of the radioactive element polonium 210, the studiedly subdued *New York Times* ran an essay written by a nuclear physicist, exploring the possibility of the substance being used in a terrorist attack. The illustration showed terrified people dying under a black rain of radioactivity. The story's subhead read: "A Russian spy dies of polonium. We should all be scared."

But the novelty of a risk does not always raise our anxiety level; nor does familiarity with it lower it. Sometimes the opposite is true—a phenomenon known to risk experts called the "availability heuristic." The term was coined by Nobel Prize–winning psychologist Daniel Kahneman, who coauthored the 1973 paper that is widely credited with inventing the field of risk psychology. Kahneman's simple but powerful insight is that the better you're able to summon up an image of a dangerous event, the likelier you are to be afraid of it. People who live in San Francisco—as Kahneman himself does—know that the risk of a catastrophic earthquake is never very far away. Nonetheless, most locals do a good job of pushing the fear down at least until an earthquake does hit. Then the anxiety over the possibility of another quake skyrockets. "The availablilty heuristic exaggerates your estimate of probability," Hahneman says. "It plants an image and that image is associated with an emotion: fear."

That fear can make us do foolish things. Kahneman is always amused by studies in which people are asked how much they would pay for a travel insurance policy that would pay them $100,000 if they died on a trip and how much they would pay for terrorism policy that would pay them the same amount if they died in an attack. Inevitably they'll pay more for terrorism protection, even though the risk is far lower and the policy thus less valuable.

Government officials are no less susceptible to this skewed view than ordinary citizens are, and that can mean very bad policy-making. John Graham, formerly the administrator of the federal Office of Information and Regulatory Affairs, routinely observed that after highly publicized incidents such as Three Mile Island, Chernobyl, or Love Canal, lawmakers and regulators were as emotional—and often as irrational—as citizens were about what steps should be taken to prevent similar things from happening in the future. "You run into this all the time within the first six months after a major disaster," he says. "But what happens is that as we get more distance from those events there's more opportunity to approach things in a reasoned way."

Individuals similarly cool down after a while, an inevitable result of the human nervous system's ability to desensitize itself to almost any unpleasant stimulus, from pain to sorrow to a persistent car alarm. But this can spell problems of its own, leading both individuals and societies to go to the other extreme in which we worry not too much but too little. September 11 and Hurricane Katrina brought calls to build impregnable walls against such tragedies ever recurring, but despite the vows, both New Orleans and the nation's security apparatus remain dangerously leaky. "People call these crises wake-up calls," says Irwin Redlener, associate director of the Mailman School of Public Health at Columbia University and director of the National Center for Disaster Preparedness, "but they're more like snooze alarms. We get agitated for a while and then we don't follow through."

We similarly misjudge risk if we feel we have some control over it, even if it's an illusory sense. The decision to drive instead of fly is the most commonly cited example, mostly because it's such a good one. Behind the wheel, we feel like the boss; in an airline seat, we might as well be cargo. White-knuckle fliers thus routinely choose the car, heedless of the fact that no more than a few hundred people die in U.S. commercial airline crashes in an average year, compared to forty-four thousand killed in highway wrecks. The most white-knuckle time of all was post–September 11, 2001, when even confident fliers took to the roads. Not surprisingly, from October through December of that year, there were one thousand more highway fatalities than in the same period the year before, at least partly because there were simply more cars around.

"It was called the 9/11 effect," says David Ropeik, independent risk consultant and a former professor at the Harvard School of Public Health. Three thousand people died as a direct result of the attacks, and a thousand more died as an indirect one.

Then too there's what Ropeik and others call "optimism bias," the phenomenon that lets us glower when we see someone driving down the street talking on a cell phone, despite the fact that we've done the very same thing, perhaps on the very same day. Of course, we were different, we tell ourselves, because our call was shorter or our business was urgent or we were able to pay attention to the road even as we talked. What optimism bias comes down to, however, is the convenient belief that risks that apply to other people don't apply to us, at least not when we don't want them to.

Even acts of nature, so we like to think, will give us a pass if it inconveniences us to think otherwise. The series of blistering heat waves that have cooked various parts of the world in recent years are a real and lethal thing. In the summer of 1995, a six-day heat wave in Chicago pushed the thermometer as high as 106 degrees during the day and only as low as the mid-eighties at night. At least 465 people died of heat-related causes during that week and a total of 521 for the entire month of July. The European heat wave of the summer of 2003 was plain shocking for its destructiveness, killing roughly 35,000 people, nearly 15,000 of them in France alone. You would assume that people would have gotten serious by now about taking precautions against such summertime perils, but that's not the case.

In 2006, professor of geography Scott Sheridan of Kent State University evaluated the heat-emergency protocols in four cities: Phoenix, Philadelphia, Toronto, and Dayton, Ohio. He focused particularly on seniors—since this population is especially endangered when the mercury rises—looking for how many of them heed such widely publicized heat-wave advice as limiting daytime activity, staying inside air-conditioned buildings, and drinking plenty of water. His finding: Not many. The reason: A lot of people over sixty-five—the threshhold heat experts use to define the top age tier of the population—just don't consider themselves seniors.

"For older people it's necessary," one sixty-five-plus respondent told Sheridan. Another added: "Heat doesn't bother me much, but I worry about my neighbors." And when denying a number as hard to refute as age wouldn't do, many seniors simply denied the numbers on the thermometer itself. Said one resident of Phoenix, a desert community where compliance was actually the lowest of all four cities: "This is cool to me; 122 degrees is hot."

Finally, and for many of us irresistibly, there's the irrational way we react to risky things that also confer some benefit. It would be a lot easier to acknowledge the perils of smoking cigarettes or eating too much ice cream if they weren't such a pleasure. It would be a lot easier never to walk into a casino if you didn't sometimes walk out with a lot of money. Drinking too much confers certain benefits too, as do risky sex, recreational drugs, and uncounted other indulgences. "These things are fun or hip, even if they can be lethal," says Ropeik. "And that pleasure is a benefit we weigh."

If these reactions are true for all of us—and they are—then all of us ought to react to risk the same way, but that's clearly not the case. Some people like roller coasters, while others won't go near them. Some people bungee jump, while others can't imagine it. Not only are thrill-seekers not put off by risk, they're actually drawn to it, seduced by a mortal frisson that horrifies other people. "There's a sort of internal thermostat that seems to control this," says risk expert and professor of geography John Adams of University College in London. "That set point varies widely from person to person and circumstance to circumstance."

No one knows how such a set point gets calibrated, but evidence suggests that it is a mix of genetic and environmental variables. In a study at the University of Delaware in 2000, researchers used personality surveys to evaluate the risk-taking behavior of 260 college students and correlated it with existing research on the brain and blood chemistry of people with thrill-seeking personalities or certain emotional disorders. Their findings support the estimate that about 40 percent of the high-thrill temperament is learned and 60 percent inherited, with telltale differences in such relevant brain chemicals as serotonin, a neurotransmitter that helps inhibit impulsive behavior and may be in short supply in people with high-wire personalities.

Given these idiosyncratic reactions, is it possible to have a rational response to risk? If we can't agree on whether something's dangerous or not, or—if it is—if it's a risk worth taking, how can we come up with laws and policies that keep all of us reasonably safe? One way ought to be to look at the numbers. Anyone can agree that a one-in-a-million risk is better than a one-in-ten, and a one-in-ten risk is better than a fifty-fifty. But things are almost always more complicated than that, a fact that corporations, politicians, and other folks with agendas to push often deftly exploit.

Take the lure of the comforting percentage—something we seem to have an especially soft spot for. In one study, Slovic found that people were more reassured by a new type of airline safety equipment if they were told that it could "potentially save 98 percent of 150 people," than if they were told it could "potentially save 150 people." On its face this makes no sense, since 98 percent of 150 people is only 147, meaning there would be three more deaths. But there was something about the seeming specificity of the number that the respondents found appealing. "Experts tend to use very analytic, mathematical tools to calculate risk," Slovic says. "The public tends to go more on their feelings."

Then too there's the art of the flawed comparison. Officials are fond of reminding the public that they face a greater risk from drowning in the bathtub, which kills 320 Americans per year, than from a new peril like mad cow disease, which has so far killed no one in the United States. That's pretty reassuring—and, as risk analysts complain, very misleading. The fact is, anyone over six years old and under eighty years old—which is to say the vast majority of the U.S. population—faces little appreciable risk at all in the tub. For most of us, the apples of drowning and the oranges of mad cow don't line up in any useful way.

But such statistical chicanery gets used all the time. Corporations defending pesticides or other toxins often argue that you stand a greater risk of being hit by a falling airplane (about 1 in 250,000 over the course of your entire life) than you do of being harmed by this or that contaminant. True enough, but if you live near an airport, the risk of being beaned is 1 in 10,000. Two very different numbers are being wrongly combined into one probability forecast. "My favorite is the one that says you stand a greater risk from dying while skydiving than you do from some pesticide," says Susan Egan Keane of the Natural Resources Defense Council. "Well I don't skydive so my risk is zero. The choices have to come from the menu of things individuals really do every day."

Risk figures can be twisted in more disastrous ways too. The 2005 book *The One Percent Doctrine,* by journalist Ron Suskind, either pleased or enraged you depending on which side of the debate over the Iraq war you took, but it hit risk analysts right where they live. The title of the book is drawn from a White House determination that if the risk of a terrorist attack in the U.S. was even 1 percent, it would be treated as if it were a 100 percent certainty. Suskind and others argue that that 1 percent possibility was never properly balanced against the tens of thousands of casualties that would unavoidably accompany a war. That's a position that admittedly became easier and easier to take over time, as Baghdad burned and the war ground on, but it's nonetheless true that a 1 percent danger that something will happen is the same as a 99 percent likelihood that it won't.

Deft—or at least cunning—policymakers can make hay out of public fears without citing numbers at all. One of the most effective devices is the art of false causation. Republican office holders running for reelection in 2006 rarely tired of reminding voters that in the five years following September 11, there was not another terrorist attack on U.S. soil, and that all of those five years were spent with the government largely in GOP hands. That's true as far as it goes, but it's also true that there was no attack on U.S. soil in the five years preceding September 11, all of them spent under a Democratic White House if not Congress. What's more, it's equally irrefutable that in the attack-free years since 2001, the New York Yankees always went to the playoffs, the Harry Potter books remained best sellers, and the potato continued to be America's favorite vegetable. Does that mean we dare not topple the Yanks, change our reading habits, or switch to carrots lest al Qaeda strike again? To be sure, the governing party deserves far greater credit for the nation's safety than such utterly unrelated things do, but any policy argument must be more compelling than merely the temporal coincidence that a particular political majority presided at a time when things were peaceful. Two facts related by time and place are not necessarily related by cause and effect.

Equally confusing—and equally susceptible to manipulation—is the devilishly complicated problem of risk versus risk. Weighing two dangers against each other, particularly when the very act of avoiding the first exposes you to the second, can freeze even the most clearheaded of us into indecision. Leap from the fifth story of a burning building and the fall might kill you; stay behind and hope for rescue and the flames could do you in just as effectively. Such mortal dilemmas are hard enough when the choices are clear ones, imposed on you by dispassionate circumstances, but things are different when it's a government or a corporation framing your options.

Adam Finkel, a professor of science, technology, and environmental policy at Princeton University, routinely delivers talks on risk, and when he does, he takes special pleasure in displaying an iconic cover from the *National Lampoon* humor magazine, one that set an early-seventies benchmark for the kind of dark satire that defined that decade. The cover's main image is a dog with its eyes photographically manipulated so they are looking nervously to its left. Entering the frame from that side of the picture is a man's hand holding a pistol. The caption reads: "If you don't buy this magazine, we'll kill this dog."

The cover may have delighted the comedic sensibilities of the college crowd to which it was intended to appeal, but it also illustrated a deeper truth of the risk equation—that being that two bad choices presented to you by someone with an interest in what you decide are almost certainly not all the options there are. "If the *Lampoon*'s threat had been a real one, you'd rightly ask, 'Well, who put the dog in that position in the first place?' " asks

Finkel. The reasonable person who does not have an investment in your buying the magazine might propose a third alternative, which is that you don't buy it *and* the dog goes free.

"You see the same thing in the real world. A dry cleaner objects to environmentalists' complaint about polluting chemicals by saying, 'OK, if you make it harder for us to clean clothes the way we do it now, we'll just have to clean them with plutonium.' But that doesn't mean that would be a real choice." As with the *Lampoon*, the dry cleaner is really interested in only one option—in this case, continuing to conduct business as usual—so the alternative is made unacceptable. But a third option, finding cleaning chemicals that are safer both than existing ones and than plutonium, is never raised.

Automakers resistant to taking on the job of designing and building fuel-efficient cars make this kind of argument in earnest all the time, reminding people that economical cars tend to be smaller and small cars are less crashworthy on the highway. Picking between the twin poisons of driving an SUV and risking a lethal crack-up, most people reasonably choose whatever it takes to keep themselves and their families safe. But the decision doesn't have to come down to twin extremes. Somewhere between the VW Beetle and the Cadillac Escalade there are, or at least could be, sturdy, well-designed cars that also go easy on gas. "It's easy for the people who control all the cards to say these are the only choices you can make," says Finkel.

So what can we do to become sharper risk handicappers? For one thing, we can take the time to learn more.

Baruch Fischoff, professor of psychology at Carnegie Mellon University, recently convened a panel of twenty communications and finance experts and asked them what they thought the likelihood of human-to human transmission of avian flu is in the next three years. They put the figure at 60 percent. He then convened a panel of twenty medical experts—who know a thing or two about the hard science of viral transmission—and asked them the same question. Their answer: 10 percent. It's the communications people, however, whose opinions we're likely to hear. "There's reason to be critical of experts," Fischoff says, "but we should not replace their judgment with laypeople's opinions."

The government must also play a role in this, finding ways to frame warnings so that people understand them. Graham says risk analysts suffer no end of headaches trying to get Americans to grasp that while nuclear power plants do pose dangers, the more imminent peril to both people and the planet comes from the toxins produced by coal-fired plants. "If you can get people to compare," he says, "then you're in a situation where you can get them to make reasoned choices."

Reminding people that they and not government officials have the final authority over those choices is equally important. The 20 to 30 percent of Americans who still don't use seat belts is an especially tragic number, because for every 1 percent you add to the population of people who do buckle up, you can save three hundred lives and prevent four thousand injuries. That's a lot of carnage avoided at a very low cost. In an attempt to boost local compliance with seat belt laws, Massachussetts made it a ticketable offense not to strap in and then, to emphasize the possible penalty, launched an ad campaign under the slogan, "Click it or ticket." It was the perfect catchphrase: clean, clear; it even rhymed. It also bombed. Say what you will about the manifest foolhardiness of failing to use seat belts, people in the United States just don't like the government telling them what to do.

"Massachusetts has a very low rate of seat belt compliance," says Ropeik. "Often, it just doesn't work to threaten people, because then they dig in their heels. Instead, you give them the information they need and let them decide." Ropeik's recommendation for an alternative slogan does just that: "Seat belts: Your choice, your life."

Easier to overlook than all of these factors—though perhaps more powerful than any of them—are the dangers that, as the risk experts put it, are hiding in plain sight. Most people no longer doubt that global warming is happening, yet we remain a nation of Hummer drivers. Most people would be far likelier to participate in a protest at a nuclear power plant than at a tobacco company, but it's smoking, not nukes, that kills twelve hundred Americans every single day. "We become so familiar with common dangers that with each passing exposure they start to seem less onerous," says McGuinn, the independent risk consultant.

We can do better, however, and political and corporate leaders can help. Indeed, this may be one area in which complexity researchers really do seem to have it nailed, in which the new science has some hard answers. The residual bits of our primitive brains may not give us any choices beyond fighting or fleeing. But the higher reasoning we've developed over millions of years gives us far greater—and far more nuanced—options. Officials who provide clear, honest numbers and a citizenry that takes the time to understand them would mean not only a smarter nation, but a safer one too.

CHAPTER EIGHT

Why is a baby the best linguist in any room?

Confused by Silence

THERE IS NO REAL REASON YOU should be able to talk. You began practicing the business of speech almost as soon as you were born, and you pretty much had it nailed before you even started school. That, however, doesn't mean it makes any sense that you can do it at all. Speech is one of the best and smartest adaptations humans have, but it's also far and away the most complicated one. Indeed, most ways you look at it, it's too complicated even to exist.

Consider the numbers: We start life entirely nonlingual,

having no conscious idea of what speech is, never mind how to understand it or reproduce it. Within eighteen months, we have a core working vocabulary of fifty words we can pronounce and a hundred or so more we understand. By the time we're three we have about a thousand words at our command and are constructing often-elaborate sentences. By our sixth birthday our vocabulary has exploded to six thousand words—meaning that we've learned, on average, three new words every day since birth. Mastering good conversational English requires a total of about fifty thousand words, and that includes only formal, dictionary-endorsed words. There are also at least fifty thousand idioms or fixed expressions—*day by day, around the block, end of the week, top of the ninth, fending off, touch and go, bundle up, buckle down, raise the roof,* and on and on. And what about children who learn a second language or a third, doubling or tripling the information that must be collected and stored and kept from commingling so that their speech doesn't simply collapse into a polyglot muddle?

However many languages we acquire, we learn to handle them surgically. We start off accumulating mostly words that describe objects, words that describe actions, and words that position those things and events in time and space. Then we learn the words that can shade the meanings of all those other words. We feel an innate difference between words like *fragile* and *delicate, sturdy* and *solid, magical* and *enchanting, sorrowful* and *melancholy,* even if we could never articulate precisely what those differences are. What's more, this is not the flukish gift of a single shaman—the way Albert Einstein's ability to part the curtain of the universe set him forever apart from the rest of us; this is one of the most democratically distributed of all human skills.

Speech acquisition may be unique in its complexity, but not in the implications of its complexity. Understanding how an extremely elaborate thing gets accomplished is always a confounding process when you take it in as a whole. It's only when you break it down to the very small things that make it go that it starts to make sense. The skyscraper is simply too big and complicated for most people to fathom fully—its size alone putting it outside what ought to be the skills of a physically tiny creature like ourselves—and yet new ones get built all the time. It takes deconstructing the building piece by piece and step by step, reverse-engineering it back to the point at which its girders were forged, its bricks were baked, the very first spade of earth got turned, to begin to get a sense of the cellular process by which such an architectural organism comes into being.

D-Day, similarly, remains the most complex military operation in history and is likely to remain so for a long time, with its 195,000 soldiers, 5,000 sea vessels, 30,000 vehicles, 13,000 parachutists, 14,000 air sorties, 13,000 tons of dropped bombs, even the 3,489 tons of soap shipped in with the soldiers for what was envisioned as a very long stay on the European continent. It's not possible even for contemporary war planners to grasp fully the scope of that, not until they run the loop backward and slowly retrace the planning, provisioning, recruiting, even the rivet-at-a-time manufacturing that made it all possible.

The same is true for a skill like speech, which we must deconstruct before we can understand how we come to do it at all. We're misled by the fact that the acquisition of the skill takes place invisibly, silently, so we conclude that the brain of the nonverbal baby somehow just awakens to the ability. The more we understand how the talent really emerges, the better we can appreciate all the pointillist processes that make up other big things. And the better we can do that, the more sense the larger universe of complexity starts to make.

THERE ARE FEW better places to begin understanding the phenomenon that is human speech than April Benasich's neuroscience lab at the Center for Behavioral and Molecular Neuroscience on the campus of Rutgers University in Newark, New Jersey. There are a lot of smart, busy people at work in Benasich's lab on any particular day, but the nimblest brains in residence usually belong to people who have no idea why they're there at all.

Take Jack, who's just over a year old. While his mother holds the boy on her lap inside a small, soundproof room, Benasich, a clinical psychologist and the director of the center, will take a measurement of his head—something she's done before, but which she must do every time he returns if she's going to keep up with his rapidly growing skull, which itself must keep up with his rapidly growing brain. Then she and her assistants will fit him with a soft, netted hat shot through with sixty-four electrodes, each of which must be moistened with conductant to ensure good contact with his scalp. One by one, the wires running from the cap must then be connected to a thick white cable running into a computer. For the moment, a cartoon playing silently on a nearby television screen keeps Jack appeased, but that's not likely to hold him long.

When the hookup is finally done, Jack and his mother are left in peace, sealed inside the little room, with even the all but inaudible buzz of the TV electromagnetically neutralized into silence. In an adjoining room, Benasich's assistants then throw a switch, and a computer-generated voice begins to speak into the chamber. It doesn't say much—just three syllables, which it repeats in random order and at various intervals: *da, ta,* and then another slightly different *ta,* flattened a bit the way it is in well-accented Spanish. To the adult ear, the sounds come way too fast and are way too similar to be remotely distinguishable. But Jack does not have an adult's ear; he has a thirteen-month-old's ear, a far sharper thing, wired to a far more complex brain.

As he watches the TV and the syllables play, his auditory center is working fast, gathering and sorting the sounds. His brain manages this task silently, invisibly, but the electrodes nonetheless detect it at work, sensing changes in what is known as the evoked response potential (ERP), electrical hiccups indicating not only that the brain's speech and hearing centers have heard the syllables, but that the differences have been distinguished. The ERP readouts of the adults in the room would be a dull, undifferentiated blur on the *da-ta-ta* test. Jack's readout is a portrait of auditory exactness.

"It's a remarkable thing to watch," says Benasich. "Hearing the differences in pronunciation like that requires you to make distinctions in sounds that take only thirty-five milliseconds to play out—and often a lot less. At that age babies not only do it well, but they do it for any sound in any language, sounds adults have long since grown entirely deaf to."

The roots of such prodigous talent begin in the design of the baby's brain. A newborn's brain looks unprepossessing, the same wrinkled, two-hemisphere mass as an adult's, but far smaller—only about as big as a large tangerine. That size differential notwithstanding, both the adult's and baby's brains contain roughly the same number of individual neurons—about 100 million—which is more or less all they'll ever have. The real difference between the two is in the wiring diagram that strings all those cells together, and here the baby has the decided edge. Any individual neuron in a baby's brain is connected to as many as fifteen thousand other neurons, and each of those fifteen thousand then branch out in fifteen thousand other ways. Across the entire body of the brain, this adds up to well more than 1 quadrillion cellular links. Adult brains have about a third fewer links per neuron, or only about ten thousand.

Elaborate as baby brains are, they get only more so after birth. In 1979, Dr. Peter Huttenlocher, a professor of pediatrics at the University of Chicago, conducted a landmark study in which he used electron microscopy to count the neurons in various brain regions of children who had died very young of accidents or diseases that did not affect brain structure. In general, he found that babies' brains indeed continue their growth after birth, but just how much growth depends on just which region of the brain is being studied. The visual cortex adds synapses until the end of the first year of life. The frontal lobes, the regions responsible for higher cognitive processes, grow through the ninth year. The temporal lobe, the area where language and hearing are processed, continues to develop until the end of the third year—precisely the age at which the child approaches some simple fluency in language.

The precise number of languages that exist in the world is something of a moving target, since they can spring into being or vanish into extinction as readily as species of animals do. In general, though, contemporary humans are thought to speak about 6,800 distinct tongues. In France alone, 75 languages—some indigenous, some not—are spoken. Tiny Papua New Guinea is home to a stunning 820. It takes about six hundred separate consonants and two hundred vowels to build such a global babble. The average individual language is composed of only about forty discrete phonemes assembled from these hundreds of sound choices, but since you can't know before birth which tongue you'll be taught, you can't know which forty phonemes you'll need. That means your brain must be versatile enough to master them all. This is one of the things that makes tackling speech so hard even before you start. Learn chemistry or mathematics or biology and the same rules apply no matter where you are. Language, by contrast, is an entirely site-specific thing, determined by the simple crapshoot of where in the world you're born.

This ability of the baby brain to run any language

program—and the speed with which it narrows its focus down to just one or a few—is one of the first things many language researchers want to understand. What they're discovering is that a great deal of early language training has to do with selective hearing. Much low comedy in the Western world has come from the inability of many Asians to distinguish between the *r* and *l* sounds. The two phonemes seem so starkly different, it's a wonder there could be any confusion at all. But the difference between the Eastern and Western ear is very real, and it appears to be inscribed in the brain.

Lotus Lin, a Tawain-born grad student at the Institute for Learning and Brain Science on the campus of the University of Washington in Seattle, has conducted studies in which Japanese and American volunteers listen to recordings of the *r* and *l* sounds being pronounced while their brains are scanned with magnetic encephelography. The images that result are evident even to lay eyes. In the American brains, sensitive to the two letters, a discrete neuronal cluster lights up, representing the precise spot at which the sound is heard and identified. In the Japanese brains there is a much larger and more diffuse splatter of activity, with neural firing beginning in the same small region and then spreading outward, as if the brain is groping around for a way to identify what it is hearing.

"We conduct test runs first," says Lin, "exposing both sets of listeners to *ba* and *wa* sounds. Since these phonemes are used in both languages, all of the subjects hear them equally well. It's only when we try the *r* and *l* that you see the difference."

This cross-cultural atonality would be a lot less broadly meaningful if it were simply a quirk of the Asian ear. But it exists everywhere language is spoken. Easterners are just as mystified by Westerners' tone deafness to the musical cadences of Cantonese. Russians are baffled at the wholly irreproducible way Italians pronounce the *gl* blend. Greeks are deaf to the diphthongs of Thais. Even among European languages, there's confusion as, say, the Spanish speaker with a tip-of-the-tongue *r* tries to move the letter to the middle of the palate where the English speaker puts it or to the back of the throat where the Frenchman prefers to keep it.

Clearly, we're not born so tin-eared, or local languages would never get learned. Perhaps the best work being done to unravel how babies calibrate their hearing to the regional tongue is being conducted by psychologist Andrew Meltzoff and speech and hearing specialist Patricia Kuhl, a husband-and-wife research team and the directors of the Seattle institute where Lin does her studies. Meltzoff and Kuhl are also two of the authors of the book *The Scientist in the Crib,* which, as its name suggests, explores the inexhaustibly inquisitive minds of newborns and toddlers.

Like Benasich's lab, Kuhn and Meltzoff's institute is at once a place of hard science and child's play, with computers and magnetic encephalographers hard up against easles, hammering toys, and shelves of board books and games. In order to study language acquisition, Kuhl devised an experiment in which a group of babies, all just a few months old, were brought to the lab for half-hour play sessions three times a week for four weeks. All of the children were from English-speaking homes and all were supervised in the sessions by a caregiver who read to them and talked to them. In some cases, the caregivers spoke only English; in other cases, they spoke only Mandarin. No single child got both languages. At the end of the month, all of the babies were brought back in and tested to see what, if anything, they had learned about the Mandarin sounds.

Since it's impossible to ask a group of prelingual children what they know and what they don't know, Kuhl devised a test in which she allowed the babies to play while a loudspeaker nearby emitted a series of random, computer-generated sounds such as *oo* and *ee*. Buried within the tones were occasional Mandarin phonemes. Babies who could make out the alien sounds tended to look toward the loudspeaker when they heard a familiar phoneme. When they did, a box lit up and a mechanical monkey inside began to play a drum. The reward was a powerful one, but the babies quickly learned that it didn't work any time they turned their heads, only if they did so in response to the Mandarin. The babies who had spent time with the English-speaking caregiver never got the reward because they never turned their heads. The other babies—exposed to just six hours of Mandarin over a full month—always responded to the sounds, performing as well as a control group of Taiwanese babies being raised on Mandarin alone.

What's more, not only did the babies exposed to Mandarin learn to hear it, they also retained that ability. When they were brought back to the lab months later—with no further Mandarin exposure—they could still distinguish the sounds they'd learned.

Impressive as those results were, the experiment was not always a success. For one thing, Kuhl found that her method worked only if the person speaking to the baby in the alien language was actually in the room. When she repeated the experiment precisely the same way but with a video image of the caregiver reading the stories and doing the cooing, the babies tuned out completely, returning the following week with no greater knowledge of Mandarin than they had before. Significantly, the video teacher works just fine if the skill being taught is not linguistic, but manual. Babies who watch a tape of an adult demonstrating how to use an unfamiliar toy will later recognize that same toy in a pile of other, unfamiliar ones and begin playing with it properly. It is only speech, with its emphasis on eye contact, matching inflections, and—especially—pointing, that requires a flesh-and-blood teacher.

"Language is an exceedingly social skill," says linguist Ray Jackendoff, codirector of the Center for Cognitive Studies at Tufts University. "What's important in teaching and learning language is that you and I attend simultaneously to the same thing, and eye contact and pointing help that happen. Remember, we're not the only species that has fingers, but we're the only one that uses them to point. There has to be a reason for that."

Age made a difference in how Kuhl's subjects performed too. On the whole, the older the babies were, the better they did at picking up Mandarin, but only until they reached nine months or so. At that point, the door to language slowly began to swing shut. Babies exposed to the new language for the first time at one year old routinely performed less well than kids half their age or younger. "Linguistically," Kuhl writes, "children start out as citizens of the world, but they don't stay that way."

Of course, new languages don't come at babies in a clear and orderly way, recited by a caregiver and rewarded with a dancing toy. Nor are they taught with constant repetitions and patient drillings the way they are in a classroom. Rather, they rain down in a meaningless stream of sounds and words. Even children growing up in a monolingual environment have a lot of sorting out to do. They may be hearing just one tongue, but it's coming at them from countless people, all of them speaking in different voices with different inflections and at different speeds. If babies are ever going to make sense of such din, they must be able to deconstruct it in real time—teasing apart its countless phonemic components as fast as they hear them. The ability to do that—and the inability that afflicts some kids—both turn on processing speed.

Words don't have to be long to be difficult; indeed, the shorter they are the more confounding they can be. A word like *people,* for example, has two syllables but only four discrete sounds—a pair of vowel sounds and a pair of consonant sounds. A word like *streams,* however, packs in five consonant sounds—with the *s* counting twice because it's pronounced two ways—and one long vowel, all carried aboard a single, clipped syllable. Benasich is currently studying how we manage to hear so many phonemes at once, focusing some of her work on what is known as the brain's flicker-fusion point.

If incoming information bombards the brain fast enough, all the data bits will eventually blur, and in some cases that's a very good thing. If you had no flicker-fusion threshold, digital music would simply be a staccato string of encoded tones instead of an unbroken flow of melody; movies would be nothing more than a jumpy series of stills, flashing at you twenty-four times every second, instead of a stream of fluid motion. When it comes to speech, however, flicker-fusion can get in the way.

To clock how fast the ear processes word bits, Benasich has babies listen to tones and phonemes separated by a three-hundred-millisecond gap—fast by most standards, but practically slow motion to the flash drive of the brain. Benasich then steadily shortens the gap, all the while monitoring the babies with ERP brain readouts. Consistently, her studies show that the babies' brains have no trouble keeping up as the space between phonemes dwindles from three hundred milliseconds to two hundred and then one hundred, eventually plunging to just thirty-five milliseconds, the point at which the length of the gap actually grows shorter than the length of the phoneme itself. Even then, the babies keep pace, getting all the way down to an eight-millisecond spread before finally going deaf to the breaks.

"The babies eventually fall behind not because they lose interest or because the brain's wiring gets overloaded," says Benasich. "Rather, it's the individual neurons that become overwhelmed. Every time a neuron hears a tone, it fires. But before it can register a new tone, it has to finish that firing and get back to chemical baseline. Eventually, the neurons simply can't do it fast enough."

Even when neurons get exhausted that way, the brain has a way to light the afterburners. Babies can somctimes process sounds spaced a mere one millisecond apart by a process known as summation. Rather than an individual brain cell transmitting a signal with its full neurochemical power, clusters of cells instead become minimally stimulated—just enough so that their collective charge can set off a few individual cells. Since none of the neurons ever has to expend much energy, they all require less time to recover, making signal processing faster. "Individual cells hit a speed limit," says Benasich. "But together, they can exceed it."

Almost all babies' brains move at more or less the same velocity, but what about those that don't? What about the ones that simply can't distinguish all the machine-gunning sounds? Benasich has found that kids with auditory-processing problems—due to either injury or a congenital wiring anomaly—tend to fall out of the flicker-fusion race at about seventy milliseconds, or roughly the speed at which the gap between phonemes is twice as long as the sounds themselves. Without brain scans, it's hard to determine which kids have such a problem—at least until the time comes that they ought to start speaking or, later, reading. Find the kids who are falling behind in these skills and conduct an ERP test, and the flicker-fusion problem may present itself. "If you can't make a precise enough phonological map of a word," says Benasich, "you can't recognize it or reproduce it."

Such a perceptual problem can be devilishly hard to

overcome. Parents understandably believe that speaking more slowly or enunciating more carefully to a child with language problems should help, if only to slow things down to the point that they can be clearly heard. But that's often useless. While it's possible to extend vowel sounds, stretching an *e* to an *eeeee* or an *oo* to an *ooooo,* consonants like *p* and *b*—the so-called plosives—can't be stretched. They come out at just a single speed and that's it. Speech pathology training and remedial reading exercises can eventually help kids compensate for processing problems, but it's a mark of the simple things on which speech stands or falls that the need for such long-term therapy can turn on just a few milliseconds of hearing either way.

"It's sometimes easier to overcome an actual injury to the speech center of the brain than it is to overcome a subtle wiring problem," Benasich says. "You can often work around a damaged area, but compensating for the way neurons are strung can be a much greater challenge."

Crisp, high-speed comprehension of sounds is just one step to mastering speech. The next is knowing how to partition those sounds further into syllables and words. Written language was a much more deliberate human invention than speech was, and so we packed it full of handy cues to keep things clear—word breaks, capital letters, periods, paragraphs. It's much easier to follow the map of a thought with white space and comma breaks pointing the way. Without such signposts, you get graphic gibberish, a stream of run-on letters that you can break apart if you must, but only slowly and with difficulty. That's how words sound to a prelingual child, or for that matter to any of us not familiar with a language we're hearing. Do you really think native speakers of Portugese or Korean or Turkish are chattering away at the gattling-gun speed they seem to be? They're not. It's your unschooled ears that are falling behind, just as theirs do when they listen to you.

The elaborately wired baby brain isn't tripped up by all this hypervelocity. One of the things it's very good at doing is quickly growing attuned to where the speakers of its native language are likely to put their word breaks; the first key to that is learning where the words are accented. In the English language, 90 percent of all words place the emphasis on the first syllable—*apple, picture, habitat, tablecloth*. Languages like Polish prefer things just the other way. Hang around speakers of one language long enough and you begin to like things the way they do. Kuhl cites brain studies in which babies less than eight months old who are being raised in English-speaking households are brought into labs and exposed to phrases that include word strings such as "the guitar is." Since the weight of the word guitar sits unfamiliarly on the second syllable, the babies' brains move the break one syllable over, correctly concluding that there is a two-syllable word in the phrase, but assuming it's *tar is*. Eventually they sort it out, but the error is no mere blunder. Rather, it's a reasonable attempt to play the language according to the rules their brains are so rapidly mastering.

Even earlier, babies seem to recognize not just syllabic inflections, but the overall direction of properly spoken speech. In 2003, Italian researchers conducted an experiment in which they scanned newborns' brains while the babies listened to recordings of speech played both backward and forward. To make it harder to distinguish which was which, they chose words or phrases that produce similar spectrographs when played in either direction, something that would make them seem at least relatively similar to the ear. Nonetheless, the babies uniformly recognized and preferred the forward-running recordings, the speech-processing regions of their left hemispheres lighting up in a much more focused way when they heard words spoken properly than when they heard them with the odd inflections and auditory swoops of inverted speech. The investigators could not be certain if the babies' brains came prewired for such a preference or if they simply grew familiar with the correct rhythms of speech very early in life—perhaps even in utero, as the muffled sounds of the outside world came in through the wall of the womb. Whatever it was, the babies clearly knew which way words ought to run and which way they oughtn't.

"The babies in the study were responding to the overall envelope of sound," says Jackendoff. "They came to the test already sensitized to attend to the beginnings of words rather than the endings, and suddenly everything was turned around on them."

Once babies start to master rhythms and word breaks, they move on to the most complex task of all: learning the grammatical architecture that assembles those sounds into thoughts, phrases, and sentences. It's no surprise that once we learn to hear words, the first ones we understand are nouns like *ball*, verbs like *run*, or descriptions like *round.* At least we can see the thing or the action being referred to. But what about a word like *of*? What about a conjunction like *and* or a future conditional like *shall* or *may*? Even now, could you clearly define an impossibly imprecise word like *would*? If you believe Murray Gell-Mann's rule that the longer the description of a thing or idea the more complicated it is, *would* would be one of the most complex words in the English language. All of these vague concepts, however, are the mortar that holds language together, and all of them are necessary to build the edifice of grammar.

Nobody knows why grammatical rules work the way they do, or why they're so similar from language to language and culture to culture. In the late 1950s, Noam Chomsky—a professor of linguistics and philosophy at the Massachusetts Institute of Technology—proposed the idea of generative, or innate, grammar. All of us, he said, are born with a set of preloaded grammatical software, basic programming that encodes the essential rules of nouns, verbs, syntax, and the rest of the linguistic tool kit. The rules are broad—adjectives needn't precede nouns in all languages; double-negatives don't always make a positive—but this very looseness allows them to govern all languages, no matter how different. Such a theory has appeal, not least because it explains so much: why certain constructions simply sound right to us and others don't; why complex phrases filled with multiple conditionals and dependent clauses are entirely comprehensible, even though we couldn't begin to explain why. It's one thing to diagram a sentence like "The blue book is on the table." It's another thing to try to unravel a syntactical tangle like "I might have acted differently if I had known what the results would be." The sentence feels right—and it is right—and that's all we need to know.

The fifty years that have elapsed since Chomsky first

proposed his theory have seen a lot of academic brickbats thrown at it—as Chomsky himself has—but none have succeeded in dinging it much. There simply isn't anything else to explain the elegant and effortless way babies learn such complex rules, particularly compared to the brute intellectual muscle—the drilling and rehearsing and memorizing of conjugation tables—we need to apply to the same job when we try to learn a language in adulthood. A 1999 study, for example, revealed that small children have an innate ability to understand the exceedingly subtle differences between such terms as *carrot-eater* and *eating carrots*—with even very young subjects correctly understanding that the first term implies a frequent act and the second implies what could be a one-time event. A 2000 study suggested that the deep rhythmic sense that allows us to feel the inherent structure of music—a sensitivity few people deny is with us from birth—may also undergird our inborn feel for the structure of grammar. This may be the reason we are not just the only species that speaks, but the only one that readily moves in time to a tune or rhythm. Jackendoff even cites research that pinpointed a precise spot along the spinal cord that appears specifically designed to control the complex breathing necessary for speech. If our spines come prerigged for elaborate language, why shouldn't our brains be programmed for the syntactical structure that makes clear communication possible?

"Deaf children who are not yet being tutored in language will begin babbling in sign, obeying rough rules of grammar," says Jackendoff. "Pidgin languages that simply rise up from other languages adhere to similar guidelines. It seems we come into the world with a natural grammatical software package."

Why then should we ever lose it? Why shouldn't the extraordinary neuronal power of the baby brain—its "exuberant growth" of synapses, as Benasich calls it—stay with us for a lifetime? If evolution indeed selects for the qualities that most improve us, it ought not select against one that would allow us to go on acquiring additional languages throughout our lives as effortlessly as we acquired the first.

One reason for our precipitous drop-off in linguistic talents, some theorists believe, might be that that kind of streamlining is simply the way complex systems behave. Corporations, committees, cities, brains—all start out with the most sprawling and varied structure they can in order to give themselves the most flexibility possible. You can't tell exactly what direction your work will take you until you actually begin doing it, so you must be able to do it all. Over time, as the system becomes more adept, it begins a process of self-consumption—a sort of refinement by reduction—digesting and shedding parts of itself in order to move toward a state of greater and greater efficiency. Physicist Norman Johnson of Los Alamos National Laboratories sees the phenomenon as almost Darwinian, with portions of the brain structure becoming vestigial—like fingers on a whale or fur on a dolphin—and simply falling away for the betterment of the whole. "In young systems, diversity serves as a sort of cannon fodder," he says. "It gets used up as function steadily improves."

In the case of language, this might be especially

important. The only way to stabilize a child's main language (or, in the case of multilinguals, the main two or three languages) is to begin hardening the brain around familiar sounds and syntaxes, filtering out distracting ones that will not be needed. Kuhl has found that, as proud as parents may be when a baby begins speaking early, the fact is the quicker babies become adept at their mother tongue, the quicker their ability to acquire other ones begins to wither.

Energy conservation plays a role in the scaling down of language ability too. Maintaining the operation of one quadrillion synapses burns a lot of calories. It's possible, some researchers believe, that the neural complexity we're born with is simply not energetically sustainable throughout our lives—particularly as we get older and the ravenous consumption of food that fuels our physical growth slows.

"We were a nutritionally marginal species early on," says psychologist William Greenough, an expert in brain development at the University of Illinois, "and a synapse is a very costly thing to support. The consequences of getting language wrong are so high that it's vitally important to get it right, which means you need a big system at first. Then you can prune it back."

Adds Jackendoff: "The thing that's really astonishing might not be that we lose so many connections, but that the brain's plasticity and growth are able to continue for as long as they do."

The language door, however, does eventually close, and the growth and learning of new languages do slow and stop. But in the long arc of a lifetime, this is not necessarily a bad thing. The command of a single preferred language that we improve and refine over decades of speaking is far more nuanced, far more lyrical than anything a child could begin to approach. *Hamlet* and *Huckelberry Finn,* after all, did not spring from the pen of a toddler. Literature, poetry, prose, even a mundane moment's conversation—all require the practice that comes only from concentrating on one, or very few, tongues. Even if we mourn the linguistic abilities we lose as we age, this focused excellence is undeniably its own triumph of complexity.

IF LANGUAGE IS OUR most elegant example of the value of capping complexity, of knowing when to pull back and trim down, there is another product of the brain that offers an equally powerful lesson in what can happen when no such limits are in place—when complexity, in effect, runs amok. You don't find it inside yourself, but all around you—in the increasingly elaborate gadgets the human mind invents. If ever there was an argument for tactical simplicity, it's in the electronic hardware that's more and more coming to govern our lives.

CHAPTER NINE

Why are your cell phone and camera so absurdly complicated?

Confused by Flexibility

IF YOU WERE FORTUNATE ENOUGH to buy a new Canon PowerShot SD400 digital camera, set aside a moment to familiarize yourself with some of the basics of taking your first picture with it. Better yet, set aside a few moments. You'll find everything you need to know between pages 61 and 111 of the 193-page manual. The section is titled "Shooting," and is helpfully broken down into twenty-five subsections, one of which is broken down further into nine subsubsections. All those mini-chapters introduce you to such things as your AE lock, your FE lock, your white balance, compression settings, AiAF, digital macro, stitch assist, and ISO speed, as well as more than thirty little icons you'll need to remember if you're going to use the right setting for the right subject. There are face, flower, noisemaker, and fish icons—for portraits, close-ups, indoor or underwater shots—as well as six variations on an eyeball and a lightning bolt for your various flash options. If that's too complicated, a flow chart and reference guide in the front of the book and a three-page table of contents will guide you through everything you need to know.

Things will be more or less the same when you open the 106-page manual for your new thirty-seven-inch Mitsubishi flat-screen TV, which will introduce you to your PC-DVI and YPbPr inputs, as well as your HDMI/DVI audio function, your timer setup and icon order menu, and your forty-three-button remote control box. Want to figure out how to use the parental lock option? Three pages of instructions and thirty-nine different keypunch commands will cover it.

Electronic devices, by any rational measure, have gone mad. It's not just your TV or your camera or your twenty-seven-button cell phone with its twenty-one different screen menus and its 124-page instruction manual. It's your camcorder and your stereo and your BlackBerry and your microwave and your dishwasher and your dryer and even your new multifunction coffeemaker, which in any sane world would have just one job to do and that's to make a good cup of coffee.

The act of buying nearly any electronic product has gone from the straightforward plug-and-play experience it used to be to a laborious, joy-killing exercise in unpacking, reading, puzzling out, configuring, testing, cursing, reconfiguring, stopping altogether to call the customer support line, then calling again an hour or two later, until you finally get whatever it is you've bought operating in some tentative configuration that more or less does all the things you want it to do—at least until some error message causes the whole precarious assembly to crash and you have to start all over again. You'll accept, as you always do, that there are countless functions that sounded vaguely interesting when you were in the store that you'll never learn to use, not to mention dozens of buttons on the front panel or remote control that you'll never touch—and you'll feel some vague sense of technophobic shame over this. Somewhere out there are people who love this stuff and are even quite good at it. If you really want to be part of the twenty-first century, oughtn't you become one of them?

Maybe—but maybe not. Increasingly, it seems more and more people have gotten fed up with the whole incomprehensible mess, and rather than struggling to accommodate the technology that's filling their lives, are demanding that the technology accommodate them—or at least insisting that somebody else figure it out. A 2006 study conducted by the Technical University of Eindhoven in the Netherlands found that consumers in the United States are now willing to struggle with a balky product for only twenty minutes before giving up and concluding it's either indecipherably complicated or simply broken. When people do throw up their hands and return the thing to the store convinced it's defective, the study found, more than half the time it's in perfect working order.

For those consumers who do put up more of a fight,

victories over technology do not come easily. A 2006 study by the J.D. Power group found that 59 percent of cell phone users have to contact their service provider for help at least once in the first twelve months they own the phone, and many call far more frequently than that. On average, it takes 1.76 calls to get a problem sorted out. That may not seem like much, but that's calls per problem—and cell phones have a lot of problems. And a 1.76 average means there are a lot of people calling three or more times for each snag they hit. The cost to consumers is more than just the time they spend on the phone, since some customer service lines—recognizing a potential bounty when they see it—charge callers by the inquiry. And those people who aren't paying up front are paying down the line. In the computer industry, the overall cost of customer support adds $95 to the cost of every new unit sold, according to a 2003 study.

If gadgets have indeed become more complex, it's in some respects unavoidable. Long for the days of the five-button television that worked straight out of the box all you want, but then you'd best be prepared to long for three channels of black-and-white programming and nothing more powerful than a pair of rabbit ear antennas to pull the broadcasts in. If you want your cable and your satellite and your DVDs and your TiVo, you're going to have to study a few more instructions and master a few more controls. The same holds true for people who lament the complexity of the computer but would be lost without the Internet, or curse their cell phone's cryptic keypad but can't imagine returning to a world in which staying in touch required a pocketful of change and a handy, working pay phone.

"Simplicity just isn't possible the way it used to be," says

Donald A. Norman, professor of design at Northwestern University and author of sixteen books on commercial technology, including *The Psychology of Everyday Things.* "You can't have hundreds of channels at your disposal and expect to control everything with a single knob. I have a single remote control that I've set up so that it handles all of my electronics, but it cost me $1,000 and I had to spend weeks programming it."

But there's necessarily complex and then there's absurdly complex. If the stripped-down gadgets of earlier generations occupy a spot on the fixed, robust end of the simplicity spectrum, don't most contemporary gizmos land far down on the chaotic side? Isn't it possible to design our gadgets so that they live at least a little closer to the top of the complexity arc, the place at which they include all of the features that most consumers would want and not a single one more? Do we even know what those features are, or have we become so fooled by our very flexibility, so confused by all of the options our technology offers, that we no longer recognize which ones do us any good? It may well be possible to create better, more intuitive, less confounding hardware, but first we must decide exactly what those devices will look like—and then we must demand them from the people who design them.

THERE ARE A LOT of things that illustrate the pleasing simplicity of electronic products from a few decades back, but few better than the way a good swift kick used to be all it took to get a sputtering television working again. Not everything that could go wrong with a TV could be set right by a kick, but if your reception was cutting in and out or your vertical hold was jumping its rails, a swat on the top or a thump on the side was often enough to bring it back in line. Now imagine trying that with your fifty-inch, high-definition plasma. Even if you could wrestle the ninety-pound thing off the wall, it's hard to imagine such a complicated bit of hardware could survive the punishment of a kick or that it would do any good anyway.

There's a reason for the simple ruggedness of yesterday's TVs compared to the fussy fragility of today's, and that is that whatever else earlier generations of televisions were, they were first of all machines. The knobs clicked on with a satisfying snap; the buttons pushed in with an audible click; the channel knob turned from position to position with a loud *chunking* sound and held that spot visibly until you turned it again, so that even if the set was off, you could look at it from across the room and know that when you switched it back on it would return to channel 2 or 4 or 7 or wherever else you'd left it.

Now look at today's TVs, with their butter-soft buttons that operate without a sound; their cradle switches for volume and channel that can be moved up and down but return, always, to the middle; their blank, implacable faces and cable boxes that reveal nothing at all when they're off. A TV that was last set to HBO with the volume up and the high-definition function on looks exactly like one that was last tuned to ESPN with the sound muted and the high-def off. "There's a lot of what we call hidden state in TVs today," says engineer Michael Bove, a director of the

Media Lab at the Massachussetts Institute of Technology. "Just by looking at the TV, you don't get any real sense of how it's operating. In the old TVs, switches really switched. Now every button simply controls a bit of software running on a microprocessor."

The honest architecture of the old televisions' exteriors was reflected in their interiors too—and this was what sometimes made a curative kick so effective. TV innards weren't made up of finely etched motherboards and microchips operating mysteriously and invisibly, but of vaccuum tubes in plug-in sockets, snaking cables, and wire harnesses. The tubes glowed a dull orange when the TV was on—causing the back of the set to give off warmth and light like a low campfire—and winked out when they died. Moving the TV from one place to another or otherwise jostling it could make any of these connections shift slightly, causing the contact to sputter or break. In the summertime—or year-round if you lived in a warm, humid place—moisture in the air could cause oxidation on the components, leading to its own kind of interruption of signals. When things got bad enough, the set would have to be taken to the shop for repairs. But most of the time, a simple, sharp smack could shake things up and reseat all the components, immediately improving performance.

"Electronics in the 1950s and 1960s were filled with things that could be plugged and unplugged," says Bove. "People simply had an intuitive sense that there was something loose inside and an equally intuitive sense that jolting it all would fix it. Now there's a hidden computer in everything, and hidden computers are hard to understand."

There is an advantage to the new electronics, of course, and that is that devices that are stuffed with computers can do a lot more things than devices that aren't. This, however, presents manufacturers with their first puzzle: How do you design a product so that users have a way to get at all those functions? One way is through the dedicated switch: Devote one control to each and every feature the apparatus has and label them all accordingly. There's no confusing a switch labeled *on* with one labeled *volume* with one labeled *channel* or *brightness*. This is the way old TVs, radios, and record players worked, and this is one of the things that made them so pleasing. Of course, old TVs, radios, and record players all had no more than half a dozen functions that would ever need controlling. You don't have to add too many more before it becomes clear that you've reached the practical limits of the dedicated switch system. The instrument panel of a 747 or a space shuttle, after all, is more or less made up of nothing but dedicated switches, but that's hardly a model for your cell phone or BlackBerry.

The only way to make the capabilities of computer-driven products fit into the palm of your hand, then, is to organize the instrument so that each button can do multiple things. The same up-and-down arrow that scrolls through your TiVo menu must also let you search for a cable movie or prowl through your prime time guide. The same few cell phone buttons that allow you to place a call must also let you text, surf, buy a ring tone, or take a picture. The key is designing the device so that people can readily—intuitively—navigate from function to function and state to state. This is where designers get involved—and that is where things start to get confusing. The fault lies less with those designers themselves than with the nature of the entire industry.

Unless you're operating a one-person business that you run from your workbench, manufacturing a product is, by definition, a collaborative enterprise. Someone imagines the thing, someone else drafts it, yet another person actually builds it, and yet another tests and markets it. "People have always liked to say that engineers build bridges," says Alan Cooper, software designer and business consultant, and author of the book *The Inmates Are Running the Asylum,* an indictment of design in the high-tech era. "But engineers don't build anything. They do the mental computing that makes the bridge possible. Drafters take over from there and, ultimately, ironworkers actually do the building."

This system has historically worked well, in part because the very scale and complexity of some jobs require different skills for different stages of the work. The engineer has the particular set of talents necessary to conceive of the bridge; but it's the drafters who figure out whether there's a better, lighter, sturdier way to do the job; and it's the workers who evaluate whether the transition from lines on a page to metal spanning a river is actually going as forecast or whether changes need to be made along the way. Put all this in the hands of the engineer alone and likely as not you'll wind up with a structural mess—one that's lovely to look at perhaps, but a mess all the same.

All that changed, however, when we moved beyond the

industrial age and into the computer age. Up until then, everything we'd built and sold existed in what designers straightforwardly think of as the world of atoms—material stuff that could be melted or burned or annealed or cut or sanded or nailed or woven or painted, transforming it from one state to another perhaps, but not changing its elemental realness. Computers, however, live in the world of electrons, at least when it comes to their software, which relies on so evanescent a thing as electrical charges flickering on and off on a microchip. That's a whole new kind of product, and one very few people can truly understand.

"A car may be made of atoms," says Cooper, "but the software that runs its navigation system is made of behavior. Software, in some ways, is just patterns of thoughts inside the heads of the designers translated to a chip. The nature of people who love electrons and bits is very different from the nature of people who love atoms."

The result has been that the computer industry has increasingly collapsed two or even three design and engineering jobs into one. The person who conceives of the software is often the same person who figures out how a user will make it work. That may streamline flow charts and workforces, but it can make a hash of the eventual product, with designers operating on the perhaps understandable but nonetheless mistaken assumption that everybody loves this stuff as much as they do, and that everybody navigates from menu to menu and command to command with the same natural ease.

It's as if the people who like experimental dance or

obscure fusion cuisine staged all the ballets and cooked all the meals and simply passed what they created on to the rest of us. That doesn't happen because in the food and dance industries there are always restaurant owners or theatrical producers who can keep the creative folks under control. Almost nobody feels competent to exert the same authority over software designers, a rarefied community of mystically gifted people who understand the black art of creating and writing code. As a result, even their bosses are unusually inclined to leave them alone, trusting them when they say that the products they're designing are the best ones possible.

"In the computer world," says Cooper, "there are two kinds of people: those who are software engineers and those who are terrified of software engineers."

This has let the designers run riot. The most obvious expression of the chaos they've wrought is not in the guts of their machines—where the software and microprocessors do their work—but in what is known as the front end, the elegant or maddening or impossible screens and menus that must be threaded through to get anything done. Anyone who remembers what it was like to work on an old personal computer before Macintosh or Windows came along remembers what it was like to try to operate machinery with almost no front end at all. Even people who never wanted to learn a thing about computers found themselves memorizing heiroglyphic-like codes and commands just to call up or copy a file. Double-clicks, icons, and pulldown menus made all that easier, and those and other conventions were quickly incorporated into all manner of other devices.

Designers, being designers, however, found it hard to leave a good thing be, and soon began producing more-elaborate front ends—the kind that only other engineers could love. The more embedded screens that could be packed into a system, the better. The more uses that can be assigned to a single key, the more functionality it seems to have.

"There's always a peer group designers are trying to impress," says Norman. "Sometimes it's the rest of the industry, sometimes it's reviewers. But these people aren't the everyday people who will be using it."

Those everyday people are paying the price. According to one survey of cell phone services, it takes an average of twenty clicks—and about two full minutes—to go online and buy a single ring tone. And that's only if you get every click right the first time. Blow a single step and the clicks and minutes multiply dramatically. As for the painstaking triple-clicking that texters have to perform to get the 5 key to type an L or the 8 key to type a P, well, there's a reason text message syntax has fractured into little more than semaphore. Closer to home, is there anyone left in the world who looks forward to navigating a DVD menu? Never mind the previews and ads front-loaded onto the disk before you actually get to the movie—those might simply be an unavoidable part of the business plan for a media company trying to make money—what about the succession of time-consuming screens and subscreens you have to slalom through before you actually get to the one that allows you to pick a scene or switch on the subtitles?

When product designers do try to keep their users' needs

in mind, they often go too far in the other direction. Arguably the most universally embraced application Microsoft ever developed is its Word program, an industry-wide standard for word processing so well thought out that even Apple—which never concedes anything to the boys in Redmond—surrendered to the inevitable and comfortably integrates it into its machines. But users accustomed to one generation of Word often face the introduction of the next generation with dread, knowing that it will be so heavily hung with new features and functions that it will become both harder to navigate and more prone to instability and crashes. And while it's true that at least a few users out there may be happy to have the Data Merge Manager or the Navigation Pane functions in their newest versions, most will probably see them as just two more distractions in a too-long pull-down menu, further camouflaging the function they really want to find. The problem is, those few users are Word customers too and the company wants to make them happy—even if it makes everyone else a little less so.

"Microsoft knows that most people are going to use only a tiny fraction of what the program can do," says Norman. "The problem is, everybody's fraction is different. Somewhere along the way, someone's asked for every item that's in the program."

The very language of software marketing has come to reflect the almost infinite mutability of almost all applications. "Look at the earliest software," says John Maeda, associate director of research of the MIT Media Lab. "The first thing you notice is that it didn't even have version numbers on it."

That telegraphed a certain settledness to the product, a solidity suggesting that what you were buying today would be the state of the art for a good long while. The ubiquitous Version 1.0 or 3.0 or 9.4 that now follows all product names communicates something else entirely—that this is a work in permanent progress, one that you'd best not get used to, because it will only be outdated next season. Part of this is the natural result of Moore's Law, the famed—and so-far fulfilled—prediction that computing power will double every eighteen months, requiring new software to keep up with new machines. But part of it is simply the result of what industry critics call feature creep, the slow, Word-style accumulation of functions that are both easy and inexpensive for talented software designers to create.

Of course, it's not just in the service of customers who ask for features that companies keep things complex. It's in the service of the balance sheet too—and few businesses know that better than your cell phone carrier. If you're like most cell phone users, your carrier probably doesn't like you all that much, because you use your phone principally—or even exclusively—to make calls. People who use their cell phone only as a phone are a little like people who pay their credit card balances in full each month. In both cases, providing the service the company nominally exists to provide is just not a moneymaker. For credit card companies, the real revenue is in the 29 percent annual interest and the increasingly usurious late fees they charge customers. (In 2006, credit card companies vaccuumed up a shocking $17.1 billion in penalty fees, a tenfold increase over the $1.7 billion they collected in 1996.)

For cell phone companies, a similar kind of supplemental profits are to be found in secondary uses like browsing the Web, mailing pictures, and buying tones or songs. Cingular (now AT&T) offered nine different categories of optional features from messaging to data to roadside assistance, ranging from a modest added cost of $1.99 per month to a more considerable $74.99, depending on the package. And none of that includes one-time charges some carriers assess every time you want to, say, e-mail a picture if you're not on a plan that covers that service. It's no wonder that cell phone companies pack their handsets with so many different ways to spend money and are so disinclined to sell you a twelve-button phone that makes a call and nothing more.

Big-box stores and other retail outlets that sell the incomprehensible gadgets the manufacturers make have incentives of their own to favor products that keep you confused. At the Consumer Electronics Show in Las Vegas in 2006, marketers discussed the significance of a newly released survey showing that fully 56 percent of people now say they would be willing to pay someone to set their electronic equipment up for them and show them how it works. Anytime so many people are willing to part with money for a service, it would be foolish to deny it to them—and the stores are not fools. Circuit City now offers its highly promoted Firedog service, a customer-assistance plan that dispatches technicians to configure and install most of the major products sold in the store, sometimes for a not-inconsiderable price. Buy a large flat-screen TV and book an installer to come in, set it up, and integrate it into your overall home entertainment system and you're parting with an additional $649.99. Add the installation of even a single supplemental speaker and you're out another $89.99. Best Buy's Geek Squad offers similar for-fee support for computers and computer networks. And while Apple does not charge customers for consultations at its in-store Genius Bars, the premium price of all Apple products certainly reflects the cost of this kind of attentiveness, and in any event, the very existence of the service brings more people through the doors, where they may be tempted to pick up more Apple gear still.

"Many electronics retailers have an understandable interest in telling you, 'The world is too complicated. We'll manage all of this for you,'" says Bove. "If a line of products came out that didn't require professional assistance, you wouldn't have to go to the store that said it had the best."

Maeda puts it more bluntly: "There's this whole business is built around complexity. It's a little like selling something dangerous and then selling the service that makes it less dangerous. But there are some things even these folks can't handle. I'm an MIT-trained computer scientist and some days I'm freaked out by my computer. If I can't fix it, I bet the Geek Squad can't fix it."

But don't blame the companies alone for keeping things muddled. The fact is, there's a part of all of us that not only doesn't mind electronics products that are daunting to use, but that actually, perversely, prefers them. The original appeal of the earliest washing machines was that they dramatically simplified what was at the time the most loathed of all household tasks. In the nineteenth century, the business of hand-washing laundry was famously labeled "the Herculean task which women all dread" by Rachel Haskell, a Nevada housewife who chronicled the hardships of wives and mothers living in Western mining camps. The mere prospect of the exhausting physical labor of cleaning an entire family's load of dirty clothes gave rise to the term Blue Monday—a reference to the sorrowful day typically set aside for the chore. If ever there was a device that called for a set-it-and-forget-it design, it was the washer, and that's precisely the kind of simple idea that long defined the machines.

But the complexity is creeping back. Temperature sensors, water analyzers, load detectors, level adjusters, touch pads, and even bacteria-killing ionizers are all being packed into washing machines courtesy of the microprocessors that make such features possible. New lights and LED readouts are proliferating on the control panels, as are the pages in the instruction manuals needed to make sense of all the features. As bad as the problem has gotten in the Unites States, it's even worse in some other parts of the world. On a tour of an appliance store in South Korea, Don Norman noticed that while the imported brands were already overly complex, the domestic ones seemed utterly unfathomable. He asked the guides escorting him through the store to explain thc difference and the answer was simple: "Because Koreans like things to look complex," they said with a shrug. An abundance of buttons is a sign of high status and quality; simplicity is a sign of shabbiness. All the same, even the most cachet-conscious consumers are likely to wind up dealing with such a complex control panel the way everyone else does.

"Ultimately," Norman says, "all those buttons are just going to drive you crazy. You'll memorize one setting and use that one all the time."

The design irony, of course, is that if any products ought to confer status, it's often not the most complex-looking ones, but the simplest-looking ones—those with their features artfully integrated into the whole rather than stamped and studded all over the outside. The iPod does a fine job of storing and playing large collections of music, but any digital player with a set of mid-priced earphones and a reasonable-sized memory could do the same. What made the little gadget a phenomenon was its extraordinary versatility. Load songs, videos, and TV shows into the memory and they're all instantly sorted for you, then offered up in a series of menus that are understandable at a glance and that you can comfortably navigate from the very first time you turn the device on. In some ways, this is anything but simplicity; it's a very complicated program doing very complicated work. But it's doing it all in the background and simply presenting you with the functional result. You don't have to think about what's going on inside the machine any more than you have to think about the fantastic mix of neural, visual, muscular, and cognitive processes that allow you to pick up a pen and sign your name. Those programs are running in the background too. Your eyes and hand are the front ends of the bodily system. The screen and scroll pad are the front ends of the iPod.

A technological world full of deep complexity and simple front ends like this would be a world full of iPods and their ilk, and that would be a very good thing. If that transformation is going to occur, it will be places like the Media Lab that will make it happen. A wholly run MIT operation since its founding twenty years ago, the lab receives additional funding from both industry and the federal government, which are supporting its mission of helping to humanize and simplify the technology both use. Over the years, newer corporate sponsors have come on board, including such companies as Motorola, Nokia, and Samsung, all makers of cell phones, and thus players with a real stake in the simplified technology game.

One of the Media Lab's most innovative pieces of hardware is something it informally calls the Bar of Soap, a name it got because that more or less captures its size and shape. Describing how the Bar of Soap looks, however, is simpler than describing what it does, since it does so many things. Like a common cell phone, it also operates as a camera, digital assistant, Web browser, e-mail hub, and, of course, a phone. Unlike a cell phone, however, it can transform itself entirely depending on how it's going to be used—much like the iPhone but more versatile still. Hold the Bar of Soap like you would hold a camera and embedded touch sensors and accelerometers signal the central processors, which paint the LCD panels that cover all sides of the unit with a digital camera's controls and screen. Hold it like you hold a phone or a Web browser and the camera controls vanish and are replaced by the ones appropriate to the new use. The genius is not in the guts of the device—the software will do whatever you tell it to do. The genius is in how it presents all those options to you. As with the iPhone, you get nothing but dedicated switches, but ones that accommodatingly disappear when you don't need them and thus conserve space.

"What you want is a device that can say 'Aha, I'm a camera right now,' or 'I'm a phone right now,'" says Bove. "It figures out for itself what it's supposed to be doing."

The Bar of Soap may be a while from making its way to market—at least at a price that more than a wealthy few could afford. But there's no reason other sensible front ends shouldn't be here already. Why should your TV make such a secret of whether it's set to cable, DVD, satellite, TiVo, or something else until you turn it on, hunt around, and navigate whatever menu map your particular configuration of systems requires you to master to switch from one input to another? Why not flash you a control panel for whatever the current setting is the moment you turn the TV on, plus a small choice of icons at the bottom of the screen that let you jump straightaway to another. "This is a way of thinking that everyone understands, but your remote control doesn't," says Bove. "The system should operate as if you were telling me to show you a movie, not as if you were telling a device."

Perhaps most important, whatever streamlined systems industry develops, it has to apply them company to company, product to product. Cell phone users currently discard and replace their units every fourteen months. If the phone you owned last year called the vibrate function the Meeting Mode and hid it inside the Profiles Menu, which was in turn embedded inside the Settings Menu, you're not going to be happy to find that your new one buries it inside a whole different set of menus and subscreens and gives it the even more cryptic name Manner Mode. Better for designers to keep in mind at all times just who's using their machines and make the products clear and consistent across the landscape of the marketplace. To the extent that companies consider these matters at all, it's often during the testing phase near the end of the production cycle, when there's only so much you can do to fix the problem.

"You don't design a good product by testing," Nelson says. "You design it by designing it well in the first place. By the time you get to testing, it's too late to fix—or at least too expensive."

If the insights of complexity science are going to be fully embraced, the world of electronics—where there are real cash profits at stake—is one place it's likely to happen soon. After all, the technology field is not the first to discover that its products have become too precious, too specialized, too much a collection of boutique goods that appeal to a handful of aficionados but leave the great mass of shoppers cold. There were only so many plates of tiny, overly styled, $60 entrees that restaurant-goers were willing to tolerate before they turned their backs on nouvelle cuisine. There are only so many black-canvas art installations people are willing to see before they start to wonder who, exactly, these things are being painted for. There are only so many cripplingly styled high-end shoes women are willing to buy before they return to wearing something they can actually walk around in.

News organizations continue to be guilty of the same kind of insularity and loss of perspective. How many newspapers and magazines now have columns or sections devoted to the media—essentially covering the people who, like them, are covering the news? Such solipsism is evident even in the arts, where there has never been a shortage of movies, books, and plays in which the lead characters are playwrights, authors, and actors. A lot of them are fine and effective narratives, but a lot of them are just exercises in self-reference, undoubtedly interesting to other people in the industry, but not to the audiences who buy the tickets.

Even noncommercial segments of society can become so caught in the swirl of their own world that they forget who it is they're supposed to be serving. There is a reason the electorate threw out the Democrats in the 1994 Congressional elections and did the same to the Republicans in 2006, and it wasn't because of courageous or controversial positions they took on substantive issues. Rather, it was because they began spending far more time on turf wars and parliamentary swordsmanship than they did tending to real business. Plenty of people thrill to Sunday morning talk shows and the drone of political debate, but plenty more just want a government that can get the trash picked up, the mail delivered, and the schools running well. Quit doing that and you're out of work.

Overly complex tech—designed mostly with other techies in mind—may have at last pushed people to that had-enough point. The good thing is, consumers will surely get what they want; in a competitive market they almost always do. It's up to the manufacturers to decide which companies will give it to them and thus still be around when the transformation—and the simplification—is done.

CHAPTER TEN

Why are only 10 percent of the world's medical resources used to treat 90 percent of its ills?

Confused by False Targets

IT'S A GOOD THING MUHAMMAD Yunus had $27 worth of takas in his pocket one afternoon in 1976. Takas don't mean much in the Western world, but in Bangladesh, they're the national currency—one that always seems to be in desperately short supply. Never were there fewer takas to go around than in 1976 when the new country, having won its independence only a few years earlier, was still finding its feet. As it turned out, Yunus's particular $27 worth went even further than most. Within thirty years, it would grow into a $6 billion global enterprise, taking

Yunus himself all the way to the 2006 Nobel Prize for Peace.

Yunus wasn't thinking about the future or prizes or anything of the kind that day in 1976. He was thinking about famine. Bangladesh is a low-lying, densely populated land located deep within the Ganges Delta—and that unfortunate topography presents some problems. The country has a long history of getting clobbered by cyclones, tornadoes, and floods, all of them alternating with the occasional drought and famine. In 1974, one of those twinned droughts and famines struck, claiming perhaps twenty-five thousand lives according to the new government's estimates—or over a million lives according to virtually every other global watchdog group without a public relations stake in keeping the mortality figure low. Whatever the toll, the Bangladeshis were clearly suffering, and Yunus, then thirty-four, wanted to see what he could do to help.

An economics professor at Bangladesh's Chittagong University, he had studied in the United States, earning his PhD at Vanderbilt University in Nashville and later joining the faculty at Middle Tennessee State, before returning to Bangladesh to teach. Not far from Yunus's home district of Chittagong was the village of Jobra, a place he had heard the suffering was especially cruel. Making the short trip out to the countryside, he found the conditions even worse than he had imagined. There was no food, little water, certainly no plumbing or electricity. Parasites and infections were everywhere, and those people who weren't dying of disease were dying of starvation. What's more, none of the villagers had the money with which they could have begun to reverse these conditions.

"As an economist, I had no tool in my toolbox to fix that kind of situation," Yunus later told an interviewer from the *Nightly Business Report* at the 2005 World Health Congress in Washington, DC. "People were dying and I felt very helpless."

Jobra, however, was not without its hopeful glimmers. Even under such oppressive circumstances, several of the locals had been able to continue practicing the various crafts that had previously earned them enough income to keep their families going. The problem was that even when they mustered the strength to produce their goods and sell them at market, creditors siezed their earnings. Yunus met a particularly talented twenty-one-year-old artisan named Sufia Begum, who wove sturdy and lovely bamboo stools but who earned no more than 2¢ per day for her troubles. The raw bamboo to make the stools cost 25¢, a quantity she needed to borrow from a moneylender, who would charge her up to 10 percent interest per week. Such a rate, she soon discovered, was unpayable. The moneylender would thus take the stools when she was finished with them, pay her the 2¢, then turn around and sell her goods for far more. "I couldn't believe someone could make only two pennies for crafting such beautiful things," Yunus told the *Business Report*.

Outraged, he decided to break the moneylender's hold. Canvassing the village, he made a list of the people—mostly women—who had a trade to practice and the willingness to work at it but lacked the little bit of money they'd need to get themselves set up. He came up with forty-two names, added up the amount each person required, and distributed the cash—$27 worth—to each villager. He imposed only one condition: that they pay him back, slowly, at the legally established market rate. The villagers agreed, and over time set up their businesses and established a small but steady income stream. They indeed paid Yunus back, plus the legal rate of interest for his troubles, and then continued to earn, unencumbered by debt—precisely the money-for-everyone way a well-executed loan should work.

That day in the village of Jobra, the Grameen Bank was born, with Yunus as its founder and director, and a simple purpose as its central mission: to make money by lending money, not in the hundreds of millions of dollars with which big banks and their big clients regularly play, but with microcredit loans, thousands and even millions of them, made to people who might not otherwise hope to be seen as a worthy credit risk. Thirty years later, Yunus the Nobel Laureate has succeeded in doing just that, building Grameen into a multinational bank that has loaned billions of dollars in its three-decade history. In Bangladesh alone there are 2,226 branches of the bank providing service in 71,371 villages, or 88 percent of all the communities nationwide, making loans that average $130, but are routinely as small as $20.

What's more, the mircocredit idea has caught on worldwide. Grameen imitators have popped up in about one hundred countries, including such oft-struggling places as India, Nepal, the Philippines, Myanmar, Vietnam, Pakistan,

Uganda, Mexico, Venezuela, and Kosovo. Worldwide, there are now more than thirty-two hundred microcredit institutions, serving over 92 million people. "Credit," Yunus is fond of saying, "is a human right."

What made Yunus's idea so ingenious and gave it its appeal is not the tiny size of his loans—anybody can hand out a small bit of money—but the precision of them. The solution to the problem of poverty, famine, and disease, he saw, was often not to carpet bomb communities or countries with money in amounts that would eventually exceed his ability to lend and in any event would not always get to the people who needed it most. Rather, the answer was surgical strikes: Find the exact point at which a little cash can do a lot of good and target your giving there. To be sure, locating that sweet spot is not an easy matter. A dying village is inevitably being killed by a lot of things at once. Do you treat the disease first? The hunger? The homelessness? The poverty? Which one lies at the root of the rest? Aim wide and your efforts—and money—will be wasted. Aim well, however, and you can crack the problem wide open.

Generations of governments and private foundations have struggled with this challenge, pouring resources into needy hot spots as diverse as Baghdad, New Orleans, sub-Saharan Africa, Eastern Europe, and America's inner cities, only to see the cash and efforts wasted and the suffering continue. In all of these cases, what's lacking is not so much good intentions or a commitment from those who are suffering, but rather a clear sense of the best spots at which to direct our efforts and the best spots to avoid. We're confused not so much by a lack of targets, but by a lot of false ones. Slowly, we're learning how to aim.

THERE ARE MANY things that make finding the weak points in the wall of global poverty and disease such a confounding thing, but perhaps the greatest is that it's one enterprise in which you can forget about any conventional concept of costs and benefits. The idea that a fixed amount of effort yields a predictable amount of result is something we take as an almost universal given, but in the arena of charitable aid that notion is turned on its head.

It was Vilfredo Pareto—a political economist born in France in the mid-nineteenth century and raised in Italy—who most clearly framed the uneven way human striving can be rewarded, when he promulgated what came to be known as his 80–20 rule. Pareto observed that in the Italian economy of his era, roughly 80 percent of the national wealth was controlled by just 20 percent of the population; the remaining 80 percent of the people were left to divide up the 20 percent of riches that remained. Once Pareto pointed out this counterintuitive imbalance, people began to notice signs of it elsewhere—the way a mere 20 percent of a journey often takes 80 percent of the time, or the final 20 percent of a project can consume 80 percent of the effort; even the way the last 20 percent of a painting or sculpture—when highlights are daubed in or contours refined—can be said to add, in some ineffable way, 80 percent of the beauty.

Profits and losses follow a kind of Pareto arithmetic too, with auto manufacturers, filmmakers, fashion designers, and others producing dozens of break-even or money-losing products, hoping for the one or two blockbusters that will tip a balance sheet from red to black. The stock market moves to Pareto rhythms as well, with a handful of turbocharged equities lifting whole sectors or a handful of laggards dragging them down.

The effort to battle disease is no exception to the Pareto rule. The National Foundation for Infantile Paralysis, better known as the March of Dimes, needed twenty-three years and tens of millions of dollars—much of it in fact raised in dime-at-a-time donations—to develop the first vaccine against polio. Once the scientists had a formula in hand, actually beating the disease was little more than a matter of firing up the factories, loading up the trucks, and getting the stuff into the arms of the children who needed it. The measles vaccine similarly took decades to develop, with investigators sifting through strain after strain of the virus before they finally found two hearty lines that could be used to immunize children against the illness without accidentally causing it. In such cases, the true sweat equity was invested during the years upon years in which only a lucky few thousand volunteers benefitted from the experimental inoculation. The later business of actually saving millions of lives took just a fraction of the work.

A close statistical cousin of Pareto's 80–20 rule is a ratio that applies only in the field of global health: the 90–10 rule. In general, for every dollar spent worldwide to battle disease, about 90¢ go to ills that threaten only 10 percent of the global population. The other 90 percent of humanity gets the remaining medical dime. From disease to disease and continent to continent this rule holds relatively steady, a fact that is simultaneously indefensible and sensible.

The indefensible part is easy enough to explain—if not to justify. Human beings, despite our pretensions otherwise, are a self-interested lot, and the people who have the money tend to spend it on the kinds of diseases that strike them and those they love. Since most of the world's wealth is concentrated in its developed nations, most of the doctoring dollars are devoted to the illnesses in those lands. Take Parkinson's disease, an undeniably cruel affliction that rightly deserves the millions of dollars spent each year in the U.S. to investigate and treat it. Nonetheless, the fortunate fact is that the number of Americans who contract the disease annually is only sixty thousand. In many respects, that's an awful lot—a stadium full of Americans year after year after year. But compare it to the stunning 500 million people worldwide who may be infected with malaria and the one hundred thousand people the disease kills every month and the Parkinson's toll looks less bleak. Yet malaria gets comparatively fleeting attention from the people with the wealth to fight it. The same kind of imbalance holds for other diseases as well: multiple sclerosis, which strikes eleven thousand Americans each year, compared to simple diarrheal diseases that have an annual global death toll of 2.2 million; Alzheimer's disease, which occurs in only 5 percent of people between sixty-five and seventy-four, versus measles, which still kills up to four hundred thousand unvaccinated victims worldwide every year—the overwhelming majority of them children under five.

Bad as those numbers are, they're even worse when you convert them into a metric called a DALY—for disability adjusted life years—which takes into consideration the total years of a healthy life lost to either disability or death. A person who ought to live to, say, seventy and is disabled or killed by a disease at forty, scores a DALY of 30 for that illness. "This is a much more thorough way of measuring the true impact of a disease," says Dr. Peter Hotez, chairman of the Department of Tropical Medicine at George Washington University and head of the Global Network for Neglected Tropical Diseases. "The DALY for malaria, which kills about 1.2 million people each year, skyrockets to 46 million. AIDS, which kills about three million people annually, has a global DALY of 84 million."

But there's more than Western self-interest behind the fact that these killers don't get nearly the cash the more selectively lethal diseases do. More compelling is the fact that, with the exception of AIDS, almost all the global scourges are diseases we already know how to prevent or cure—and cheaply too. A measles vaccination costs 16¢ a shot; treatment for often-lethal intestinal worms goes for 2¢. Treatment for parasitic river blindness costs $1; polio vaccine goes for 16¢. Even bed nets, which protect people against malaria-carrying mosquitos, cost no more than $5 each. And while that comes out to $2.5 billion to cover just the people in the world who already have the disease, it's only a tiny fraction of what it costs to treat and hospitalize and, ultimately, bury so many of them. The illnesses that preoccupy the West don't require merely a pill or shot or simple intervention, but batteries of repeated, expensive medications and treatments—not to mention costly tests and extended hospital stays. And none of that includes the millions upon millions of dollars spent in labs each year in search of better cures or therapies.

Looked at that way, it's the developed world that's actually in worse shape than the Third World, if only because simple, well-targeted solutions to Western ills just don't exist yet. There are few things that better capture the power of easy fixes to solve crises in the Third World than the battle against cholera and other diarrheal diseases. And there are few things that better capture the beginning of that battle than the 1962 epidemic in Dhaka, the capital of what was then East Pakistan.

The worst that can be said about diarrhea in the West is that it's an uncomfortable nuisance, one that usually clears up in a day or two. In the developing world it's a pitiless killer. Of the five leading causes of death among children under five in 2006, diarrheal disease was number two, killing 1.9 million babies and preschoolers. Malaria, a mass killer itself, trailed far behind at 853,000. The only disease more lethal than those two was pneumonia, with 2 million children under five claimed.

What makes diarrhea so deadly is not the one hundred or so viruses, bacteria, and parasites that cause it, but the body's attempt to get rid of them, which it does by violently purging itself of anything that enters the gut. A body spilling its contents that way can't possibly absorb any nutrients or, more important, water. And with up to

75 percent of the body made of nothing *but* water, that can be deadly. Victims of diarrhea-related dehydration can die in less than a day, as fever climbs, the kidneys quit, and shock sets in. Babies, with the least to spare in terms of strength and immunity, are the first to go.

"People suffering from cholera can lose about a quart of fluid an hour," says Dr. William B. Greenough, a professor of medicine and international health at Johns Hopkins University. "Altogether, you can lose about five gallons in a day. There's a reason untreated cholera is up to 40 percent fatal."

At the time of the 1962 outbreak, Greenough was a young physician working for the National Institutes of Health and was dispatched to Dhaka as part of an international team to see if it was possible to stop the dying. When the doctors arrived, they were taken straight to a jute mill, a place usually used to manufacture industrial fibers, but now being used to warehouse the dying. At least five thousand refugees were crowded into the place, many of whom would not last the day. "The only thing we could do was separate the sicker ones from the healthier ones and try to save those we could," Greenough recalls. "That meant rehydrating them."

But rehydration wasn't so easy to do. At the time, the only way to get water into the body and ensure that it would not pass straight through was to introduce it intravenously, about twenty liters per day per patient. But even if there had been enough transfusion fluid available, IVs are slow, complicated, and require a lot of equipment—little of which was on-hand in a jute mill. The answer was to figure out a way to bypass the needle and make oral rehydration work.

And the way to do that was to come up with something to add to the water so that it would stay in the body long enough to do some good.

Nobody cracked that secret in time to save the victims of the 1962 epidemic, and true to cholera's cruel arithmetic, four out of every ten people struck by that outbreak died. Not long after, however, the answer to the rehydration problem—an astonishingly simple one—was found. Never mind a lot of complicated chemistry, investigators found that merely by mixing a pint of clean water mixed with about a handful of sugar and a three-finger pinch of salt—though precision is critical lest you actually exacerbate the dehydration—they could produce a fluid that the choleric body can ingest and hold. The sugar, which is easily absorbed into the intestinal lining, begins the recovery cycle, changing the chemistry of the cells in the gut in such a way that they become more receptive to taking up the salt. Salt, in turn, helps the system retain water, and water saves lives. It was nine years before another cholera epidemic struck Dhaka. That time, health care workers were ready.

"In the 1971 epidemic," says Greenough, "the rehydration fluid let us slash the death rate from 40 percent to just 3 percent. When we supplemented the worst cases with IV rehydration, we could cut that 3 percent to practically nothing." The salt-sugar-water mixture became the accepted protocol from that moment on, and today, the precise formula has been so well refined that more complex salts and sugars can be used—virtually eliminating the risks associated with the preparation—sealed into packets and shipped to the site of an outbreak, where they merely need to be mixed with water.

The cost of a packet sufficient for a liter is just 6¢, and numerous groups like the World Health Organization and UNICEF are making the treatments increasingly available.

So why do 1.9 million children keep dying each year? First of all, that figure, awful as it is, is actually a lot better than in the prehydration days, when 5 million were claimed annually. "Ask what else has saved that many lives in the past thirty years and the answer is nothing," says Greenough. Second, just because an effective rehydration method has been created doesn't mean the delivery infrastructure exists to get it to the people who need it.

More problematic, rehydration is useful only in one part of the diarrheal cycle—the part after a person is already sick. A better strategy is to keep victims from falling ill in the first place. Governments and private groups in Africa are now teaching villagers the simple art of building a pit toilet, something that costs almost nothing at all, but sequesters disease-carrying waste and cuts infection rates 50 percent. In one Ethiopian district, the number of people with access to some kind of latrine facility rose from 3 percent to 86 percent in just three years, with pit toilets even becoming something of a status symbol. Another cheap and simple answer is common zinc tablets, which help protect the intestines and cut relapse rates 30 percent in recovering patients at a cost of as little as 25¢ for a ten-day course. Also now available is a newly developed vaccine against rotavirus, the leading cause of diarrheal diseases. A dose of the preventive sells for a head-spinning $60 in the West but goes for no more than $7 per treatment in poor countries. And that figure is likely to come down further as competing vaccines are developed and charitable foundations negotiate discounts.

"Bad hygiene and bad water have always been the biggest problems driving these diseases," says Greenough. "Those have hardly been eliminated, but it's been an extraordinary success by groups like the W.H.O. and UNICEF to get us this far."

Even more extraordinary work is being done by the Carter Center in Atlanta, partly because it's being done by such a comparatively small group of people. Former President Jimmy Carter has devoted much of his multi-decade postpresidency to battling diseases in the developing world, but he's never had even a small fraction of the staff of a global group like W.H.O., nor is he sure he'd want one if he could have one.

"We have only 150 people on staff to cover the entire world," says Carter, "and that includes the groundskeepers at the Carter Center in Atlanta." What makes such a small team possible is the advantage of Carter being Carter. Occasional controversies aside, he remains deeply admired throughout most of the world, and that means his calls always get returned.

"If UNICEF or W.H.O. wants to go into a country and discuss local disease conditions, they would hire ten people and start by working with the minister of health," Carter says. "I don't have to do that. I send word in advance to the king or the prime minister, to the ministers of health and finance and agriculture, the whole cabinet, and say we want to send someone in."

Carter has made particularly effective use of this clout in

his battle against guinea worm. A painful, disfiguring, and often lethal parasite, guinea worm is passed from person to person via contaminated drinking water. One way to beat the disease is simply to clean up the rivers, ponds, and other water sources with larvacides that kill the parasites. Another way—less comprehensive, perhaps, but still effective—is for locals to use fine-mesh cloth to filter water thoroughly before it's consumed. The parasites are large enough to be trapped by the weave, making it safe—or at least safer—for locals to drink the water. The problem is, the filtration cloths tend to rot, as constant contact with water combines with the hot, humid climate to cause the fabric to dissolve over time.

During one of Carter's trips to Africa in the late 1980s he observed the filtration practice in action and resolved to help work out the rotting problem and popularize the simple safety measure throughout the guinea worm–afflicted parts of the continent. Upon his return to the United States, he paid a call on businessman Edgar Bronfman, whose family owned the Seagrams company and had just acquired a large stake in Dupont.

As Carter and Bronfman sat down to lunch, the former president made the purpose of his visit quickly clear, explaining the ravages of guinea worm, the number of people afflicted, and the comparative ease with which it could be wiped out. Then he plucked up a cloth napkin from the table and told Bronfman that that was essentially all it would take to end the crisis. Shortly afterward, Bronfman contacted the Dupont board of directors, proposing that the company develop a rot-resistant filter fabric that could be used against the disease. Dupont agreed, developed the cloth, and donated 6 million square meters of it to the Carter Center. Since 1986, the incidence of guinea worm in Africa—which at that time affected 3.5 million people—has been reduced a remarkable 99 percent. That's by no means all a result of the filter cloths; education and cleanup programs and the development of better treatments and larvacides have played an enormous role too. But it was the filters that began it all.

OFFERING NEW TREATMENTS or teaching locals new skills may be the most obvious ways to stop an epidemic, but they're not the only ones. Often the solution is simply a matter of delivery—developing innovative ways to get an existing cure or vaccine to people who have no access to it. Perhaps the greatest epidemiological success story of all time is the eradication of smallpox. That victory, historic as it was, wound up turning on a very small thing—just millimeters worth of small, in fact.

It was in 1796 that English physician Edward Jenner first noticed something surprising about smallpox. While the virus could strike pretty much anyone pretty much any time, the disease did seem to avoid one small substratum of British society: milkmaids. Even during rural epidemics, the disease would invade countrysides and farms, but oddly stop short when it reached the girls who worked with the cows.

Jenner began investigating the mystery and soon concluded that the reason the milkmaids weren't getting smallpox was that in a small way, they'd all had it already. A close cousin of the smallpox virus is the cowpox virus, which, as its name suggests, afflicts principally cattle. Milkmaids who spent hours each day with their hands on livestock couldn't help but pick up a case of the disease themselves, often repeatedly, though it was almost always a mild one since the virus wasn't adapted to humans. Still, the similar-but-not-identical infection was enough to stir the girls' immune systems into action, producing antibodies that would protect them from a broad array of pox viruses later on. Jenner built on this insight, developing a way to collect virus samples from sickened cows and expose humans to them through a small scratch on the skin. The technique worked and the first vaccine was born.

It was more than 150 years before refined formulations of Jenner's potion went into wide use around the world. By the 1960s, the disease was all but eradicated—at least in the West. In Africa and other developing lands, however, smallpox continued to rage. Finally, the W.H.O., the U.S. Centers for Disease Control, and the Peace Corps decided it was time to wipe out the virus for good and launched a worldwide vaccination program. But a couple of obstacles stood in the way.

For one thing, the smallpox vaccine, like many vaccines, required constant refrigeration. That's not so hard in the developed world, where insulated boxes and refrigerated trucks make it easy to maintain what scientists call a cold chain, over which perishables can be carried without spoiling. Doing the same in the developing world, where electricity, refrigerators, and even trucks and roads are not always in reliable supply, was a knottier problem. Just as vexing was the fact that even if you could get cold, fresh vaccine where it needed to go, there was no guarantee it could be administered properly. Unlike, say, the oral polio vaccine, which almost anyone can serve up with little more than an eyedropper, the smallpox vaccine requires an injection and a particularly well-trained hand to administer it. The needle can't be plunged too deeply into the muscle or the vaccine will be pulled apart by the currents of the bloodstream before it can do any good. And it can't be inserted too shallowly or the formula will never be absorbed.

"We had a vaccine that worked just fine," says Dr. Donald Hopkins, a veteran physician who worked with the Centers for Disease Control at the time and now advises the Carter Center. "But in too many cases it wasn't getting to people in a way that did them any good."

The answer to the cold-chain problem was relatively easy: Freeze-dry the vaccine and rehydrate it on-site just before it's used. The answer to the administration problem required a more elegant solution. Instead of relying on an ordinary syringe, designers developed one with a short, forked needle, with the two prongs spaced a tiny distance apart. The reduced length ensured that the points of the needle would be long enough to break the skin to just the right depth and no more. The forked design meant that all you had to do was insert the needle into the vaccine and surface tension would hold a precise measure of the liquid between the points like a drop of medicinal dew. Then it was just a matter of jabbing the skin several times and the solution would penetrate to the right spot in the right dose. The needle worked perfectly and was rolled out with the international eradication campaign on January 1, 1967. By 1977, a village in Somalia reported the last case of wild smallpox anywhere in the world. "A very tiny thing," says Hopkins, "led to a very big result."

IF LOW-COST MICRO-SOLUTIONS like this are possible, why does it still cost so much to keep the world healthy? Part of the problem is that spending money and spending it well are often two different things, and governments—which traditionally have the most cash—are usually the least skilled at knowing what to do with it. The United States has what is arguably the best-funded health infrastructure in the world, and yet we're still a nation staggering under epidemics of obesity, cancer, heart disease, AIDS, and even malnutrition. But look at where the health money goes. The total budget request of the U.S. Department of Health and Human Services in 2007 was $74 billion. With that money, however, the department must support a workforce of 67,000 employees spread out among seven major divisions, including the departments of Medicare and Medicaid services. Among the many expenditures from the overall pot of HHS cash that have nothing to do with actually keeping people healthy and making sick ones well is a $1.4 billion allocation to run the department's Program Support Center, a bureaucratic arm to provide "customer-focused administrative services and products." Whatever that is, it doesn't sound like it's saving a lot of lives.

Washington has routinely been guilty of this kind of profligate if well-intentioned spending. Remember the 1976 swine flu vaccine—a $450 million national program to inoculate Americans against a disease that never really struck? Remember the millions of dollars spent on the antibiotic Cipro during the 2001 anthrax panic—with the U.S. government alone plunking down $100 million for 100 million capsules? In both cases, a great deal of money was spent and almost nothing meaningful was accomplished. On the other hand, early in the early 1980s, when the AIDS epidemic was just beginning to roar to life, the Centers for Disease Control asked Washington for $30 million for an HIV prevention campaign and was sent packing. Certainly, there was little available at the time that could have prolonged the lives of those already infected. But how many other lives—and how much more money—might have been saved if that $30 million had been allocated and then imaginatively spent to help people avoid infection in the first place?

Private foundations and charitable groups, which ought to be nimbler, if only because they're smaller, are often little better at getting value for their money. Shippers, bookkeepers, manufacturers, bureaucrats, warehouses, truckers, and others all get a chance to touch the cash or medicines an organization is trying to move from one place to another, and even if every one of them remains scrupulously honest, a lot still gets lost along the way in salaries, fees, purchase costs, and other expenses. Add the always-present risk of corruption, which is endemic everywhere, but particularly in the Third World, and even the best-planned and best-funded efforts can vanish into what health experts ruefully call M&Ms—mansions and Mercedeses—for corrupt local officials. Over the decades, so much money and medical treasure has vanished in war-torn parts of Africa that givers simply began despairing of the whole place and shutting off the grants altogether—one reason the continent remains as ill as it is today.

Lately, private foundations have begun figuring out how to avoid this kind of waste, and as they have, they have begun returning to Africa. The biggest of all the charitable groups—and the one with arguably the most exacting standards for efficiency—is the Bill and Melinda Gates Foundation. Even for an organization so huge, the Gateses know that the first and most important virtue is frugality. The foundation's total available capital exceeded $32 billion in 2006, and while this is small by the standards of a government, it's mammoth by the standards of a private charity. Nonetheless, in any one year, the trustees can give away no more than about 5 percent of that total, since all they can really safely spend is the revenues from the investments in which the overall endowment is parked. Spend any more than that and you eat your seed corn and quickly kill your fund. Even in a good year like 2006, that means the Gateses could makes grants totaling no more than $1.3 billion—or about the cost of running the Iraq War for a week. But you don't found a company like Microsoft without learning a few things about maximizing resources.

For one thing, the Gates Foundation has set things up so it doesn't get blindsided by events. All manner of wealth is wasted when grantors ship foods or medicine to needy countries only to find that legal details block the supplies at the airport, poor roads prevent shipments from being transported inland, or a key medicine that was intended for one substrain of a virus is being used to battle another. Before grants are issued, Gates overseers analyze the local conditions to spot these problems in time to sidestep them. Sometimes this requires them to know only about the local politics or infrastructure. Other times it requires an exhaustive understanding of microbiology, even by people not trained in the field. Bill Gates himself has often done just this, once famously learning to be a lay expert on hookworm, a parasitic disease that kills tens of thousands of people each year.

"Of the helminths, I am not sure of which ones rely on human hosting as key to maintaining the life cycle," he once wrote his scientific adviser in a letter reprinted in *The Chronicle of Philanthropy*, a trade journal. "If it is key, then you have a chance of dramatic reduction if you block the life cycle in people."

This non–medical man's intrusion into the work of doctors may exasperate some of the experts, but it also pushes them to imagine problems first and then avoid them. Helen Gayle, a Gates Foundation grant director, once worked for the federal Centers for Disease Control and spent enough time answering questions on Capitol Hill to know a thing or two about being grilled. Still, she was unprepared for the exactitude of the Gateses. "Bill and Melinda are as tough as any Congressional panel I've faced," she told the *Chronicle*.

Once a grant is made, the foundation stays stubbornly

close to the recipients. The Gateses are especially aggressive when it comes to monitoring school grants. Of the three dozen Gates employees who work in the foundation's education arm, at least nine are assigned exclusively to the job of tracking the performance of grant recipients. At one Gates-supported school, the foundation displayed its knack for spotting where a trouble spot is when it backed a plan to remove all of the doors from the school building except those leading to the bathrooms and the outdoors—the thinking being that this would foster a better atmosphere of communal quiet and serious learning. Such a radical recommendation would not work in many—actually most—schools, but that's just the point. In this one, the analyses suggested it would—and it did.

The most important strategy, however, one practiced not just by the Gateses but by a growing number of health foundations, is the business of bundling drugs and services together so that a single medical dollar does several jobs at once. The Global Network for Neglected Childhood Diseases is attacking seven different parasitic ills endangering people in fifty-six different countries, and doing so with a single envelope of multipurpose pills that costs just 50¢ per pack. The Carter Center, hoping to develop infrastructure in Africa that can provide a pipeline through which multiple drugs or treatments can flow, has helped develop village-based distribution networks flexible enough to dispense whatever medicine or service is needed for whatever epidemics that may arise. The Gateses are attacking the problem of infant mortality—which annually kills 4 million babies under one year old—not by trying to battle each of the countless diseases that cause the dying, but with carefully monitored grants to other foundations, like the Save the Children Foundation, which in turn are providing basic health information to 20 million mothers and training thirteen thousand health care workers to treat them.

"Sometimes, you're teaching something as simple as the need for skin-to-skin contact between baby and mother—what we call kangaroo care," says Mary Taylor, a Gates Foundation program coordinator. "Yes, it's often necessary to deliver a drug or a service. But other times you're just teaching behaviors."

As the global health network slowly awakens to the wisdom of such flexibility, precision, and planning, ever more imaginative ideas are saving ever greater numbers of lives. In Africa, a privately sponsored group called Riders for Health found an unexploited weak spot in local medical-care programs in the lack of agile vehicles that could get needed medicines and doctors to villages isolated by washed-out or nonexistent roads. So with backing from Save the Children and local governments, they began servicing and purchasing as many new and secondhand ambulances, refrigerated trucks, and motorcycles as their resources would allow and providing them to local health care agencies. This motorized fleet now ranges over great stretches of Uganda, Gambia, Somalia, Ghana, and other nations in Africa. In Zimbabwe alone, one thousand of the vehicles now service the entire 13 million-person country, slashing mortality rates from some diseases—including malaria—by 20 percent.

In Botswana, where 300,000 people out of a population of just 1.6 million are infected with HIV, Ernst Darkoh, a

Harvard MBA who was born of Ghanian parents and reared in Tanzania and Kenya, determined that the best way to put the brakes on the epidemic was to introduce a little market savvy into the health care sector. He recruited Western distribution experts to streamline commercial delivery routes in order to get goods quickly to market and, using the time saved by this more efficient delivery, then put the same trucks to work in the more urgent taks of getting drugs and treatments into the field. He changed the procedure at health care clinics so that patients who came in for any treatment also automatically got an HIV test unless they specifically requested not to have one. Most effectively, he introduced the equivalent of express lanes at crowded AIDS clinics—a technique he learned from studying Wal-Mart—so that people who only needed medication or testing got fast-tracked to one team of health care providers, and sicker patients could receive more time-consuming care from others.

In Nepal, government health officials who were having difficulty coaxing families to supplement their children's diets with vitamin A—a problem that was contributing to a shocking infant mortality rate of 133 out of every 1,000 live births—decided that what was needed was people to distribute the pills who had not only the time to get involved in volunteer work but the moral authority to ensure the kids jolly well took the vitamins that were handed to them. The answer: the country's underutilized population of grandmothers. Today, there are 49,000 volunteer grandmothers helping to distribute vitamin A to 3.5 million Nepalese children. Since the 1980s, infant mortality has been cut in half.

Even Mohammed Yunus's already finely targeted microloans are being refined further. The Grameen Bank recently began expanding its reach, launching what has become known as its beggars' program, providing loans beyond the ordinary poor and out to the wholly destitute, people who had despaired of even trying anymore. The program now makes such applicants eligible to receive advances as tiny as $9, backstopped if necessary by local communities or private guarantors. For these desperate people, this may be just enough to provide a little food and medicine and a place to sleep. That in turn provides them the stability to try to reenter the workforce—even if that work is little more than a variation on the door-to-door panhandling they do anyway.

"We say to beggars, 'Look, as you go house to house, would you carry some merchandise with you—some cookies, candies, toys for the kids—to sell?'" Yunus said at the World Health Congress. The beggars have begun carrying the goods, and the people do sometimes offer to buy. What's more, before they hand over their money, they offer their visitors—beggars no more, but now respectable peddlers—something entirely new to them. "As the beggars show their merchandise, they are given a stool to sit on, which they never had before," says Yunus. "They not only sell, but they get respect from the families."

IF THE MERE act of offering a beggar a place to sit can be such a curative thing—and for the beggar, it can be—there ought to be no lingering reason for the wealthier, more resource-rich world to continue insisting that it can't help the more infirm world. Yes, it's far too simplistic to say that with enough commitment and resolve we can wipe out poverty and disease. The shape-shifting nature of viruses and bacteria will always provide them with new avenues of attack, just as the ever-changing global economy will always leave some people fatally behind.

But it's also too simplistic to say that the full-blown crisis in global health can't be brought to an end—and in comparatively easy and affordable ways. We know too much about the enormous power of little things—the way just the right tap can split a gem, or a ball hitting the sweet spot of a racket can sail off seemingly forever—to pretend that the $9 loan and the 6¢ vaccine and all such little fixes can't also have a power that greatly belies their size. Our deep sense that this is an effort we not only could be making but should be making comes from a place that might, ultimately, be one of the most uncomplicated parts of us: our simple sense of compassion.

CHAPTER ELEVEN

Why does complexity science fall flat in the arts?

Confused by Loveliness

BYRON JANIS WAS NOT ONE TO ignore the advice of Vladimir Horowitz—not in 1947 anyway. Janis was a nineteen-year-old piano prodigy, stupendously gifted and enormously promising, but still years away from the global fame he would eventually enjoy. Horowitz, by contrast, was Horowitz, already equal parts performer and monument, and arguably the most gifted expatriate pianist Russia had ever produced.

Horowitz had first heard Janis perform four years earlier, when the then fifteen-year-old made his network radio debut with the NBC orchestra. Conductor Arturo Toscanini heard the performance too, phoned Horowitz, and suggested that so talented a boy might benefit from a tutor. Horowitz, who had never taught before, agreed, and for the next several years, the master coached the student, sometimes on the magnificent piano in Horowitz's Manhattan apartment, sometimes at Toscanini's home in Riverdale, New York. No matter where the lessons took place, Janis came to accept that an instruction from Horowitz was not a thing to be dismissed lightly. So when the occasion arose that he considered doing just that, he surprised even himself.

The suggestion from Horowitz that Janis ignored concerned the way to play Maurice Ravel's composition *Jeaux d'Eaux,* or The Fountain. Horowitz told Janis that Ravel himself insisted the piece be played entirely without pedal—an odd bit of dictum for a delicate piece that occasionally required a jolt of extra power. *Jeaux d'Eaux* without pedal in the sound seemed like it would be a feeble thing indeed. Janis nonetheless tried it as he'd been instructed, and the result was as weak as he'd feared it would be—what he thought of as a dry fountain. It was not the kind of thing he'd want to be caught performing.

"I decided not to listen to him on this one," Janis says today. "From that point and for thirty years after, I always played the piece the way I thought it should be played."

What continued to mystify Janis, however, was why Ravel would have handed down such a rule in the first place. The answer came in the late 1970s, when Janis was on a performance tour of Europe and decided to visit the late composer's home just outside of Paris. Ravel's house and studio were places many pianists came to pay homage and, if they were sufficiently well regarded, to try out the Bechstein grand on which Ravel had composed so many of his great works. Janis was one of those invited to sit and play. He chose *Jeaux d'Eaux* and, in a gesture that was either brave or insolent, went at it with full pedal. It sounded awful—oppressive, aggressive, a tone wholly wrong for the delicacy of the piece. When he eased off the pedal, the beauty and balance returned. Janis knew his Bechsteins and knew it couldn't be the nature of the piano or the quality of its tuning that was responsible for the sound.

What was to blame was nothing more than the size of the room. Ravel composed in a small study just twelve feet wide and twenty-four feet long, a modest space that may have helped him shut out the world as he worked, but did nothing to give full range to his piano's sounds. Pedal, in this little room, would blow out the walls. Composers, a headstrong lot by nature, have a stubborn tendency to fall in love with a piece the way it sounded when they first heard it, and Ravel was no exception. If pedal was too much in the study or the parlor, it must be too much in the concert hall as well. On such occasions, it would be left to other, later performers to work out how even a great master's best compositions ought to be performed. "Music depends on discipline," Janis says. "But it also depends on freedom."

Grand art doesn't always stand or fall on such small

details, but often it does. In the engineer's world, things can more or less work. In the economist's world, a theory can sometimes apply. But the artist's world is a place less tolerant of almosts and half successes. A painting either transports you or it doesn't; a piano concerto either soars or stays earthbound. Often, it's impossible for even the creators of art themselves to know what makes the difference. And if the artists can't tell, what chance do outside observers, least of all scientists, have?

No matter how powerful the insights of simplicity and complexity theory have become, they have often stopped at art's edge—and well they should. You might be able to get at the basic bits of optical or auditory science that make our greatest masterpieces what they are, but suggesting that is what explains the zing of the whole is like saying cogs explain the clock or DNA explains the tiger. Those are the parts; the sorcery is in the thing they become.

One reason complexity theory hasn't trod much in art's yard is that scientists, for all their chilly literalism, seem to know this. Louis Armstrong spoke a deep truth when he said, "If you have to ask what jazz is, you'll never know." And while lay music fans may disagree, convinced they can peel apart an Armstrong riff and grasp by sheer brain power what their gut isn't telling them, accomplished musicians understand Armstrong's wisdom. So too, oddly, might scientists. The Santa Fe Institute lists twenty-six different disciplines it explores, from areas as broad as the origin and synthesis of life to those as specialized as cryptography, and yet it wisely touches art only glancingly. We can be sufficiently fooled by the power of pure loveliness to believe that there are equally powerful answers to be found if only we dig deeply enough. To be sure, the sheer scope of both the world of complexity and the world of art do ensure that on occasion the two overlap. The key is to know what those convergences mean—and what they don't.

OF ALL THE artistic disciplines complexity researchers might explore, it is perhaps music that best lends itself to investigation. Music, like science, is a field with its own coded language, its own rhythmic rules, even its own way of drawing up what amount to equations, in the form of sheet music and written scores. That's the kind of stuff scientists love. What might be the biggest news in the relatively little-known field of musical science broke in 2007, when the journal *Science* published a study by Dmitri Tymoczko, a composer, music theorist, and assistant professor at Princeton University—marking the first time a musicology paper had appeared in the journal in its 127-year history. It took something ambitious to catch the publication's eye, and Tymoczko provided it.

Western musicologists are very good at studying Western forms of music, largely because they make auditory sense to Western ears. The chords we consider pleasing are typically made of notes separated by at least one step and rarely more than three. Notes that bump up against each other like two adjacent keys on a piano produce what we take to be a clashing sound. Contemporary music gives these so-called cluster chords a chance, but in a limited way. Not only do we allow ourselves a relatively fixed number of chords, we build them into melodies in an equally fixed way. Rather than jumping all over the scale, we keep the distance between any two sequential chords as small as possible. A tune we perceive as pleasant may eventually wander a fair distance from the vicinity of the first chords struck, but it takes it a while and it does so only gradually. Music from other cultures that strays too far from these rules seems to us to be deconstructed and arbitrary.

"For the most part," says Tymoczko, "the music we like satisfies two basic constraints: It has chords—collections of simultaneously appearing notes—and these chords are articulated by independent melodies, which move by short distances as the chords change."

Unfamiliar music is not only hard for us to hear, it's hard for us to diagram, with the familiar five-line staff not always suited to the job, particularly when the music is mostly percussive. Relying on the staff as our sole musical guide is a little like having an excellent street map of our own city, without ever getting a look at a global atlas that would show where the place we know best fits. Tymoczko wanted to see if there might be a way to map that larger world.

The first thing he did in pursuit of that goal was to blow up the five-line staff and replace it with something known as an orbifold, a mathematical construct borrowed from the fields of physics and economics. Drawn on a piece of paper, the orbifold looks like a sort of conical prism, one that folds back on itself the way a figure eight does, at least mathematically. Mapping harmony and counterpoint within this flexible, three-dimensional space provides a far fuller picture of how musical forms fit together, in much the way a globe reveals that traveling very far east will eventually bring you very far west—a fact a flat map entirely conceals.

Tymoczko mathematically modeled dozens of musical forms and instructed his computer to position them around the orbifold. To his surprise, the potentially chaotic universe that resulted appeared to be a comparatively ordered place. All music we consider mainstream, he discovered, moves around in a very limited space near the center of the orbifold, obeying the rules of incremental chord progression. Seemingly deconstructed forms of music live in different parts of the prism, but similarly don't stray very far from their home turf, abiding by their own laws of harmonic chords and conservative changes. We may not hear the order in those forms, but the computer shows it exists.

"These constraints are satisfied by a large variety of Western music, stretching back many centuries," Tymoczko says. "They are also satisfied by a lot of Western-influenced non-Western music and some kinds of atonal music." Further refinements of the orbifold model will likely gather in even more forms of music, particularly those from Africa, which Tymoczko finds particularly interesting and so far have not been assigned coordinates on his map.

The question, of course, is what, if anything, the new model means. Tymoczko himself stresses that while such a spatial understanding of melody and counterpoint might help musicologists and laypeople better grasp music theory, it would be largely useless to a composer writing music—at least one without a deep knowledge of physics and computers. Even for that rare person, the orbifold would serve mostly as a tool—a little like a sharper pencil or a better piano—something that could make the work of composing easier, but in no event make average music good, much less good music great.

What's more, it's not even clear just how new any finding like this is. It's never been a secret that music has a complex internal arithmetic—one that mirrors the larger arithmetic of nature as a whole. Much of basic mathematics, for example, pivots on the idea of balance—of the quantity on the left side of a formula being the same as the quantity on the right. Young pianists are taught to think in similar zero-sum terms, as Janis recalls from his own youth. The left hand is the anchor hand, the one that grounds a performance. The right hand is something in flight. The right can do whatever it wants as long as the left essentially pays its rhythmic debt. It's this idea that gave rise to the term *tempo rubato*—Italian for stolen time—which refers to the way an added note cannot simply be dropped into a musical phrase, but rather must be taken from a point later in that phrase, lest the whole composition become unstable.

Everywhere in nature this kind of equilibrium is maintained, as all matter and energy search for the relaxation pathways that keep systems balanced. It's the reason air rushes in to fill a vacuum, the reason ice melts and warms toward the temperature of the room, the reason structures under pressure collapse to relieve the stress. It even governs the way spacecraft on their way to a distant planet will slingshot around a closer one in order to benefit from a sort of gravitational whip crack that adds speed and gets them where they're going sooner. The energy boost seems to come free, but actually there's a bill to be paid—and it's the planet itself that covers the tab. Even as the tiny spacecraft accelerates by thousands of miles per hour, the giant world slows down by a fraction of a fraction of a millimeter. In such a way are the ledgers of physics kept balanced—and in such a way is all of nature's bookkeeping done. To think that a musician composing a score is aware of any of this is to misunderstand art. But to think such rules don't lay in the bedrock of the art is to misunderstand science.

"Bach's music has always been thought of as very exacting, even mathematical," says Janis. "But his original compositions are filled with the most emotional, impassioned handwriting. There is a connection between mathematics and music, but it's the emotion that comes first."

Music is hardly the only kind of art that abides by a powerful underlying order. So do forms that appear to be nothing but disorder. Take the seemingly chaotic works of Jackson Pollock. Few people beyond the painter's most hardened critics argue that his drip-and-splatter masterpieces are entirely without structure. But they're hardly works of pointillist precision either. What made the paintings remarkable—and even transformative—was Pollock's improbable union of the free-form exuberance that is so plain to see and an iron discipline that was often entirely hidden. For that, he owed at least some debt to physics.

The art world crackled in 2006 with the news that six possible Pollocks whose authenticity had never been established were being examined by physicist Richard Taylor of the University of Oregon in an attempt to determine their origin. In the 1990s, Taylor conducted extensive examinations of a number of known Pollocks, superimposing a computerized grid over the canvases and analyzing the patterns he found inside each square. This helped him determine the stylistic signature that linked them all. He found even more than he bargained for. Not only were the paintings not as arbitrary as detractors suggested, they actually obeyed the complex law of fractals—the tendency of natural patterns to repeat themselves at smaller and smaller scales, the way the branching of watery tributaries mirror the larger branching of river channels and the channels in turn mirror the arms of the river; or the way the hourly fluctuations of the stock market resemble the daily, monthly, and yearly changes.

Pollock's seemingly uncontrolled patterns were found to repeat themselves at smaller scales too, though he clearly never set out with a series of finer and finer brushes to produce such duplication deliberately. Rather, it was the physics of his muscles as he poured paint onto his canvases that was likely responsible. Biomechanical analyses reveal that as the off-balance body regains its stability it behaves fractally as well, with orientation changing and musculature flexing in repeating ways at different scales. This is not just an attribute of artists. Every time you skid on a throw rug and recover your footing before you fall is an exercise in fractal motion too. What distinguished Pollock from the rest of us was the way he could regulate this—intuitively more than consciously—controlling not only the larger patterns that formed as his fractally scattered paints fell, but where those patterns would land: The canvases themselves were filled with fractal signatures, but the drop cloths that surrounded them, Taylor found, were covered by nothing but meaningless splatters.

When Taylor subjected the six contested paintings to his computer scrutiny in 2006, he found that they, like the drop cloths, lacked the signature fractals that would have marked them as true Pollocks. What the scientist had on his hands was not necessarily frauds, but rather the works of one of the artist's many imitators and followers. What they manifestly weren't were the paintings of the master.

"Pollock was honing his fractals a quarter of a century before they were defined," wrote the editors of the journal *Nature* when the Taylor studies were released. Taylor himself put it more simply. "Pollock was in control," he said.

Just as works that appear chaotic can have hidden structure, works that appear clean and disciplined can be governed by even more rigidly controlled design rules than they seem to be. Some of the world's most painstakingly assembled art is found in the mosaic tilings on the walls of Islamic structures built in the Middle Ages. The complex, seemingly symmetrical geometry that makes up the works had always been assumed to be built around the idea of pattern repetition, with particular configurations of tile clusters appearing again and again, side by side, to form a larger, symmetrical whole. Not only did the works look like they were based on such redundancy, simple mathematics dictated that in most cases they must be: When a broadly symmetrical mosaic gets large enough—as these are—it becomes arithmetically very difficult for the internal patterns not to repeat.

In the 1980s, however, mathematicians proved that nature could produce what's known as quasicrystals, so named because they form seemingly symmetrical patterns that expand indefinitely but—on minutely close examination—never repeat themselves. Such things seemed neither physically nor numerically possible, and yet the natural world was manufacturing them all along. As it later turned out, so was the Islamic world. In 2007, a Harvard physicist and a Princeton cosmologist took another look at mosaics in Afghanistan, Iran, Iraq, and Turkey and found, to their surprise, that their patterns were quasicrystalline too. The key to the ancient designs was in the individual stars and polygons that make up the mosaics, which can be rotated at varying degrees and locked into neighboring stars and polygons in varying ways. That adds just enough variety to the growing pattern that a configuration of tiles that appears to be repetitive isn't. Adding to the subtle variety are colorful lines of tile that slash through the mosaic at mathematically predictable but nonrepeating angles. Those two bits of inspiration made them perfect, fifteenth-century expressions of the quasicrystal form.

What was driving the long-ago artists to marry such high craft with high physics is impossible to know. Paul Steinhardt, the Princeton cosmologist behind the discovery, did not even attempt to guess how medieval designers could reproduce patterns that twentieth-century scientists believed they had discovered. "I can just say what's on the walls," he told John Noble Wilford of *The New York Times*. Clearly, though, the tilers began with at least an innate sense of how the universe works, even if the tools did not yet exist to spell it out scientifically. The same must have been true of Pollock, who never wrote or spoke of fractal-like patterns, but created them all the same. And the same must also have been true for the celebrated designer Buckminster Fuller.

Fuller is most famous for his cunning geodesic dome, a self-supporting structure, made up of pentagons and hexagons, that requires no central support beams and actually becomes stronger and stabler the larger it grows—at least up to a point. It was in 1949 that Fuller built his first dome, but it wasn't until the late 1980s that chemists proved such things exist in nature too, in the form of spherical carbon molecules that form in such unremarkable places as candle soot and precisely follow Fuller's pentagon-hexagon structure. The molecules too become stronger and stabler up to a certain size—in this case sixty carbon atoms—but far smaller and far larger ones exist as well. The chemists, who later won the Nobel Prize for their discovery, named the molecules buckminsterfullerenes or, more colloquially, buckyballs, a tribute to the designer who intuited that such structures existed—or that if they didn't, they ought to—and brought them forth decades before science could catch up with him.

IT IS BY no means the modern generation of complexity scientists alone who have searched for hidden patterns in the abstract world of art. It was that hunt for structure, after all, that drove early twentieth-century Harvard linguist George Zipf to discover his eponymous Zipf's Law, the curious way the most frequently occurring word in any text will appear roughly twice as often as the second most common one, three times as often as the third, and so on.

In 1886, a similar interest in linguistic patterns drove American physicist Thomas Mendenhall to try to settle the seemingly eternal debate over whether some of William Shakespeare's plays were in fact written by Francis Bacon, by comparing the frequency and distribution of short, four-letter words in the text of the two men. Around the same time, Polish philosopher Wicenty Lutosławski attempted something more ambitious. In trying to determine the sequence in which Plato wrote his celebrated dialogues, he defined no fewer than five hundred stylistic features in the ancient philosopher's work and devised a formula for tallying and ranking them in significance. Since the work of any artist evolves, he reasoned, any one writing sample that's statistically—and thus stylistically—very close to another should have been produced close in time as well.

"You simply draw a graph in which the vertical axis is

the percentage of an author's texts in which a certain stylistic profile appears, and the horizontal axis is the year each was written," says Dan Rockmore, professor of mathematics and computer science at Dartmouth College, and an external faculty member of the Santa Fe Institute specializing in the statistics of style. "This will give you a line of some kind. Any text from that author that you later stumble on will fit somewhere along that line." It's unknown if that's precisely how Lutoslawski did his work, but his elaborate analysis could certainly be reduced to such a straightforward schematic.

The greatest of the mathematical text analysts, however, was undoubtedly the nineteenth-century Russian mathematician Andrei Markov, whose self-named Markov models were so useful to political scientists in understanding the conditions that preceded the fall of Apartheid and the Berlin Wall. Even more powerful Markov models provide a statistical measure that calculates the frequencies of certain phonemes and words and uses this to predict the other words likeliest to surround them. In a body of colloquial text, "knick" is almost 100 percent certain to be followed by "knack." The term "a dozen" is not quite as likely to be followed by "eggs," but it's still a fair bet. In any paragraph in which the words "add," "subtract," and "multiply" appear, there is at least an increased likelihood that you'll find "divide" too.

"One of the first things Markov did with his model was analyze the cadences in Aleksandr Pushkin's novel in verse *Eugene Onegin,*" says Rockmore. "This didn't make anyone look at Pushkin's work differently, but if you're following the road of Markov chains, that's where it starts." Markov modeling thrives today as a powerful tool to study physics and chemistry and drive speech-recognition software.

Similar Markovian and non-Markovian computer programs can even play a role in more sensational cases, such as efforts to unmask literary forgeries or expose authors who write under noms de plume. Political reporter Joe Klein published his 1996 novel *Primary Colors*—a roman à clef of Bill Clinton's 1992 presidential campaign—under the name "Anonymous," and for a short while he was able to maintain that cover. A lot of clues pointed to him, not least being his reportorial familiarity with the Clinton campaign, but it was ultimately a computer analysis conducted by Vassar College English professor Donald Foster that pinpointed small, idiosyncratic details in Klein's writing that forced him out of the shadows. Just six months after the publication of the best-selling book, Klein held a press conference to concede that he was the author.

A similar reduction of style to data points was applied in the field of painting long before fractal analysis came along. In the nineteenth century, Italian art critic Giovanni Morelli developed a method for determining the provenance of paintings by hunting for details as seemingly subjective as the folds in the curtains in a painting of a room or the shape of earlobes in a portrait. This gave rise to the popular Morellian method of analyzing and authenticating art and is the first approach both professionals and amateurs use today. "Morellian analysis is precisely what a docent leading a tour group through a museum will do," says Rockmore, "pointing out the flows here or the fleshy shapes there. In a deeper and more complex way, it's what art curators do all the time."

Coaches, physiologists, and biomechanical experts increasingly apply similarly detailed analyses to the movements of athletes and dancers in an effort to help them improve their performances. Originally, this was done simply by eyeballing them as they went through their paces and spotting what there was to see. Later, film allowed for a closer, frame-by-frame analysis. In the past twenty-five years, this has given way to ever more sophisticated digitization of motion, with computerized body sensors and superimposed grids allowing a much more fine-grained examination of form. The technique has even found a place in moviemaking and television production, as human motion is captured and digitized in greater and greater detail, so that the performances of living actors can more believably be translated into cartoons, commercials, and animated films.

IN ALL OF this literalizing of art, there's a wonderful—even liberating—uselessness, a simple why-would-you-bother question that's hard to answer. And it's here that the still young, very broad field of complexity research may reach a limit. Understanding the hard science of a thing is not always the same as being able to make any use of that knowledge. Does a dancer or gymnast who masters the classroom study of gravitational physics learn a thing about how better to control a leap or stick a landing? Does the painter who studies the frequencies of the visible spectrum have an edge over one who doesn't? Does a psychologist who can tell you why you laughed at a joke illuminate or annihilate the humor?

All art, no matter who produces it, will be slave to larger guiding rules, but in the end, that will tell you nothing about the unquantifiable loveliness of the thing. Soaring literature will display the same Zipf distribution as leaden literature; transcendant music lives in the same orbifold as carnival music. Complexity science doesn't even do you any good when you try to make sense of the rarest of works—those produced by artists at the pinnacle of true greatness. Conduct all the empirical analyses you want, and you're still not going to learn what separates the Bacon from the Shakespeare, the Salieri from the Mozart, the physical gifts of the ordinary athlete from the sweet swings and balletic stride of a Willy Mays or a Walter Payton.

None of that, of course, has kept scientists from trying, from acting as if simply understanding the parts of something is the same as being able to assemble them into a whole. Artificial flavors are built literally molecule by molecule, with food scientists reducing a strawberry or vanilla bean or some other food to vapor, running that essence through a gas chromatograph and blueprinting the molecule chemistry. Then it's a simple matter of recreating and mass producing those molecules artificially. But does this really fool anybody? Is there anyone with an even nominally sensitive palate who can't distinguish between a piece of fruit grown on a vine and its chemical doppelgänger cooked in a lab? Motion-capture technology may produce cartoon characters and computer-generated bodies that are more anatomically fluid than ever, but to a movie audience squirming through a film like the unintentionally creepy *Polar Express,* the experience is something badly lacking. People avoided the movie not because of the writing or the performances, but because of the expressionless eyes and vampiric lifelessness of the faces. Genuine human voice plus genuine human motion does not equal genuine humans—or even an engaging facsimile. It equals something that is eerily other.

SMART COMPLEXITY RESEARCHERS recognize this, knowing that there are whole landscapes for them to explore without treading in places where theory and the scientific method avail them nothing. Byron Janis once made this point in a conversation he had with James Watson, the Nobel Prize–winning codiscoverer of DNA. The two became friendly after a performance Janis gave at Watson's Cold Spring Harbor Laboratory in New York, but it wasn't long before they realized that while they occupied the same social world, they lived in very different professional ones.

"Success in science," the pianist recalls telling the laureate, "depends on repetition. You must do everything thirty-five times or it doesn't count. Success in the arts depends on nonrepetition. You try something unusual and maybe it works, but you never try to repeat it, because then you'll destroy it."

And yet, every now and then, the two worlds—the artistic and the scientific, the ineffable and the empirical—do find a sweet spot, a place they flow together in a rare moment of convergence. Janis experienced such a bit of serendipity one night in 1957 when he was recording a performance with the Chicago Symphony Orchestra. Like all such recordings, this one could not be conducted with an actual audience in the house, since random coughs and other acoustical shocks are impossible to control. Eliminating the audience, however, also comes at a price. A mass of people of different heights, girths, and clothing types provides something of an auditory sweetener, a soft surface that both absorbs and reflects sound waves, but in any event improves and enriches them. The risk of errant noise, however, is greater than the payoff, so the house is usually kept clear.

Janis sat down to play that evening and was immediately unhappy with what he heard; the sound was chilly and thin and all wrong for the piece. But he thought he had an answer. Backstage, he had noticed a long piece of plywood, about eighteen inches taller than the height of the piano. He asked a stagehand to bring it to him, then carefully leaned it up against the left side of the piano, so that it rose out above the lowest base keys, forming a sort of angled awning over them. Wood, Janis knew, is a good sound reflector, and this particular piece might reflect in an especially happy way.

He began to play, and as he'd suspected, all of the missing depth and richness of the sound returned. The lifeless piece of wood, positioned just so, did the acoustical job of an audience full of living listeners. Janis was not the man to tease out all the molecular, vibrational, perceptual, and architectural science that made that happen. But the people who could answer those questions were not the ones to know intuitively what a mere piece of plywood could do in that moment, or to play the music that that insight allowed. It was in the collision—the collaboration—of those two worlds that the music lived that night. That's a place too complex to be understood, and too simple not to be.

Epilogue

MURRAY GELL-MANN IS NOT SOMEone who denies himself his vanities. It's hard to win a Nobel Prize when you're barely out of your thirties and not let it go to your head a little bit. And even now, years later, Gell-Mann is clearly a man of strong opinions who appears to suffer fools only grudgingly. Still, his most particular vanity is actually a rather disarming one.

Having devoted more than twenty years of his life to studying the riddles of simplicity and complexity, Gell-Mann has come to see them as such interwoven parts of a whole that he likes to refer to them with his own little shorthand word: plectics. It comes from the root word *plexus,* or folded, with simple meaning folded once and complex meaning folded numerous times. It's indeed a neat neologism, and Gell-Mann shows it off with a whiff of the pride he no doubt allowed himself when he introduced the world to his subatomic quarks so many years ago.

"I invented the word," Gell-Mann says simply. "It's very well thought out."

It's a measure, perhaps, of the precision involved in studying complexity that even so seemingly frivolous a concern as nomenclature can be a cause for such consideration. Some things really ought to be too precise, too *fine-grained* as the scientists at Gell-Mann's Santa Fe Institute like to put it, to warrant your attention. Still, the trick for people studying complexity and simplicity remains a question of focal point—knowing when you're looking too closely and when you're not. A newpaper picture is a complex thing at a distance, but a simple thing up close, when it resolves itself down to nothing but a collection of black dots, all more or less the same except for their size. But aren't there dimensions of complexity inside the composition of the ink and the absorptive science that causes it to cling to the paper? Didn't Gell-Mann's Nobel itself turn on the question of focal point, of looking at the once irreducible atomic nucleus and reducing it much further? Wisdom comes in knowing when that closer look is worth the effort and when it isn't.

As complexity science matures, it is likely to get better and better at mastering that question of approach. For

Gell-Mann, the most elegant example of the power of such perspective occurred long ago, with the work of a scientist who came along well before him: Karl Jansky, the early twentieth-century physicist who founded the science of radio astronomy.

In 1931, Jansky, then twenty-six, was working for Bell Telephone Laboratories in New Jersey, when he was asked to solve a stubborn problem. Overseas telephone reception, still in its relative infancy, was often washed out by a steady and distracting hiss of static that even the best equipment and newest lines could not eliminate. Jansky and others knew what some of the source of the noise might be. Lightning, for example, was certainly part of the problem; but those disruptions ought to come mostly in bursts, occuring more or less simultaneously with each flash in the sky. This constant crackle was something else entirely.

Jansky set about trying to solve the riddle and built a large directional antenna that he mounted on a set of wheels from a Model T Ford, allowing him to aim toward various spots in the sky and determine at least the general direction from which the noise was coming. Rotating his antenna around and around over the course of months, he ruled out atmospheric interference, since humidity, temperature, and other variables of the air did not seem to affect the noise. More disappointingly, he ruled out incoming energy from the sun, a good candidate since, like the static on the line, it was always present. But solar energy has its own cycles to which the interference didn't correspond.

What Jansky did notice was that the noise peaked every twenty-three hours and fifty-six minutes, or the exact length of time it takes the Earth to spin on its axis once. And at the precise moment the interference was the most powerful, his antenna was always pointed toward the constellation Saggitarius—or more or less the center of the Milky Way. The noise, he realized, was coming from the very heart of the galaxy itself, from powerful radio waves emitted by stars, cosmic clouds, and all the other agglomerations of matter that make up the local cosmos.

Up until that moment, visible light—just a tiny band of the broad electromagnetic spectrum that includes x-rays, gamma rays, microwaves, and more—had been the only way to study space, a little like standing three steps away from a door and trying to determine what's on the other side by squinting at the keyhole. Jansky threw the door open, allowing a flood of new information in and giving rise to much of the great astronomy that's been performed since then. For that achievement, he's an icon not just to the astronomical community, but to complexity researchers too.

What Jansky did, after all, was gaze into the chaos of a cosmic storm and find something deeply complex there, something Gell-Mann calls "beautiful regularity." The coarse-grained cosmos became suddenly fine-grained under Jansky's eye. While it's mostly scientists who will ever appreciate the magnitude of that achievement, all of us have the chance to consider it every day. Even as phone lines have improved and global communications have grown clearer, we can still listen for Jansky's lovely noise in the quiet crackling lingering in the background of an ordinary radio broadcast. The song may be the sound you want to hear and the static may be the noise you don't. But the true key to grasping complexity is to understand the music in both.

Author's Notes

THE CREATION OF ANY BOOK IS AN exercise in often daunting complexity. For helping me bring it under some kind of control, I owe a lot of people a lot of thanks.

My greatest appreciation, of course, goes to the folks at the Santa Fe Institute, including Samuel Bowles, George Cowan, Jennifer Dunne, J. Doyne Farmer, Jessica Flack, Michelle Girvan, Ellen Goldberg, Betty Korber, David Krakauer, John Miller, Ole Peters, D. Eric Smith, Constantin Tsallis, John Wilkins, and C. C. Wood. Thanks too to Susan Ballati for making it possible for me to speak to all of these gifted people. And thanks most of all to Murray Gell-Mann and Geoffrey West, the men at the helm of the Santa Fe ship.

Numerous other scientists and experts aided me in my work, including David Axelrod, Robert Axelrod, April Benasich, Michael Bove, Michael Brasky, Norm Chow, Alan Cooper, Michael Coy, Lisa Curran, William Greenough of the University of Illinois, William B. Greenough of Johns Hopkins University, Brooke Harrington, Donald Hopkins, Ray Jackendoff, Norman Johnson, Azalea Kim, Steven Lansing, Blake LeBaron, Simon Levin, Lotus Lin, Anna Lingeris, Ken Massey, Donald Norman, Dean Oliver, Scott Page, David Romer, Scott Ruthfield, Sam Schwartz, Carey Stever, Sander van der Leeuw, and John Wertheim. I am indebted as well to Alison Gopnik, Andrew Meltzoff, and Patricia K. Kuhl, authors of the fine book *The Scientist in the Crib,* which gave me an early introduction to the wonder of infant speech. Special thanks to Byron Janis and to President Jimmy Carter for their time and extraordinary insight.

Chapter Two ("Why is it so hard to leave a burning building or an endangered city?") is expanded from a cover story I originally wrote for *Time* magazine. Thanks to the folks at *Time* for permitting its repurposing here. Even deeper thanks to Dan Cray in the Los Angeles bureau and David Bjerklie in New York for the always extraordinary reporting they contributed to the story.

Background and reporting for Chapter Ten ("Why are only 10 percent of the world's resources used to treat 90 percent of its ills?") was also drawn from stories I and others wrote or reported for *Time* magazine. Thanks particularly to Andrea Gerlin, for her remarkable cover story in *Time* Asia on diarrheal diseases, as well as Robert Horn, Alex Perry, and Simon Robinson, who filed to the story. Similar appreciation to Megin Lindow and Michael Brunton for their splendid reporting in *Time*'s 2005 global health issue.

More personally, thanks as always to the remarkable Joy Harris of the Joy Harris Literary Agency, who liked the idea for this book and early in the process called to tell me that I ought to call it *Simplexity,* a wonderful neologism that occurred to her at three o'clock in the morning. Thanks too to Will Schwalbe of Hyperion Books for helping me get my arms around so big a topic and giving it the shape and structure it needed. Appreciation as well goes to *Time* science writer Michael D. Lemonick, who, upon first hearing the idea for this book, pointed west and recommended I get myself to Santa Fe as fast as I could manage.

Finally, as always, thanks and deep love to Alejandra, Elisa, and Paloma, for helping it all make surpassing sense.

Index

Bobbie Buch

JEFFREY KLUGER is a senior editor and writer for *Time* magazine. With astronaut Jim Lovell, he wrote *Apollo 13*, on which the 1995 movie was based. His other books include the critically acclaimed *Splendid Solution: Jonas Salk and the Conquest of Polio*. Kluger lives in New York City with his wife and daughters.